W9-BZU-161

CONTENTS

Log Homes Made Easy

Log Homes
Made Easy

Contracting and Building
Your Own Log Home

Jim Cooper

STACKPOLE
BOOKS

Published by
STACKPOLE BOOKS
Cameron and Kelker Streets
P.O. Box 1831
Harrisburg, PA 17105

Printed in the United States of America

10 9 8 7 6

First Edition
Cover design by Caroline Miller with Christine Mercer
Cover photo by Jim Cooper
Interior line drawings by Michael Musto

Library of Congress Cataloging-in-Publication Data

Cooper, Jim, 1949–
    Log homes made easy : contracting and building your
own log home /
Jim Cooper.
        p.   cm.
    ISBN 0-8117-2422-0
    1. Log cabins — Design and construction. 2. Contractors.
I. Title.
TH4840.C66   1993
690'.837 — dc20   92-32751
                                                  CIP

Introduction

FIRE FLICKERS IN a stone hearth, sending glimmers of orange and scarlet dancing across walls of solid wood. Overhead shadows play among massive beams. Outside, snow sifts down through bare limbs to blanket the ground, while inside on the sofa, the family cat curls itself tighter. The aroma of home cooking hangs in the halls, mingling with a fragrance of wood and the faint scent of the fireplace.

This is log home living. It is a dream pursued by many. And for many, the dream will remain just that, like learning to play the piano or sailing solo around the world. But each year, more and more people are bringing their log home fantasies to life. Indeed, statistics suggest that more than one out of twenty custom homes built today is built from logs, a doubling from five years ago.

The lure of a home set away from the sprawl of the suburbs, nestled in tranquil woods or perched on a rural hillside, is not hard to fathom. As the end of the twentieth century approaches and people find themselves more and more caught by the rush of technology, their lives surrounded by the products of laboratories and factories, the pull of something natural and peaceful grows stronger.

For many, log homes represent a haven from technology, a reminder of simpler, quieter times, and a link to a rich and rugged heritage. Even

The appearance and "feel" of a log home are affected greatly by the log and corner style. Here, large diameter Swedish-coped logs are joined at a notched corner with the log overhangs scribe-cut into graceful curves. Finish details such as the custom cedar door contribute to an overall atmosphere of rustic elegance.

though today's log home is a masterpiece of modern engineering and sophisticated technology, the overwhelming presence of wood makes it seem like a page out of history.

But the dream of log home living carries a real price tag. As with anything "custom" today, costs can be higher. In addition, because of the uniqueness of log homes and the bewildering array of choices, purchasing and completing one involves more research, planning, effort, and determination than purchasing a conventional home. The procedure is further complicated by a second dream that often goes along with owning a log home — the dream of building it yourself.

Once upon a time, log homes were tiny, owner-built cabins in forest clearings. When the owners finished their cabins, their children studied law by firelight and grew up to become presidents. Times have changed. The modern log home industry began as a source for inexpensive second homes — weekend retreats at the lake, in the woods, or in the mountains. Small cabins were (and still are) well within the means of weekend carpenters and amateur builders. But the weekend log retreat led many to decide that it would be fun to live in a log home all year. Somewhere over the last decade, log second homes have given way to log primary residences. Today the majority of log homes are being built for full-time occupancy.

With the shift from cabin to year-round home, log homes have gotten larger and more sophisticated. What works in the woods on weekends is not what most people want to come home to every night. But the desire for a bigger, more sophisticated residence has not always brought the recognition that building a home is far more complex than building a weekend cabin.

I'm always amazed when someone comes through my door with that "do-it-yourselfer" glow and unrolls a set of plans that would give pause to a good builder of conventional homes — huge homes with fourteen corners or more, cathedral ceilings everywhere, dormers on top of dormers. This person would never consider attempting such a project using conventional framing, but for some reason the "simplicity" of stacking logs makes the whole idea seem downright easy. Never mind that logs account only for the exterior walls, leaving a very complicated roof to frame and a substantial amount of interior framing.

On a large house project, do-it-yourselfers face the potential for economic disaster. Not only are the risks of mis-estimating (this means underestimating) enormous, opportunities for structural catastrophe are magnified. Underestimating construction costs by 10 percent on a project

Glass-fronted living areas are popular, adding a contemporary counterpoint to the logs. By carefully considering placement of large glass exposures, homeowners can use solar energy to enhance the energy efficiency of log walls.

expected to cost less than $80,000 is survivable by most people. But a 10 percent error on a $200,000 project can produce gruesome results. A 10 percent margin of error is not uncommon among owner-builders, and twice that is not unrealistic.

While I don't want to discourage owner-builders completely, I do want to sound a strong warning. I've heard of too many log home dreams shattered because enthusiasm overpowered common sense. If your building experience is limited, your project is a primary residence comparable in size to most conventional residences in your area, and your principal reason for building yourself is to hold costs down, you have already taken three giant steps toward disaster. If you believe that you can build your log home for less than a tract-built conventional home of similar size and design, disaster is almost assured.

Fortunately, the increasing number and sophistication of log homes is causing people to recognize their limitations when it comes to construction. Fewer people have the time, energy, or inclination to go out and set logs themselves. But the desire to be involved in the building process remains strong. So today the owner-builder is giving way to the owner-contractor.

This book is written for the owner-contractor—the individual who has no intention of taking a chainsaw or broadaxe into the woods to forge a new life in the wilderness. I write for the individual or the couple who wants the satisfaction of bringing their log home dream to life under their direction. They will probably have to balance their contractor responsibilities with jobs and family. They will be people who recognize that even though they may save money acting as their own contractor, that savings will be on the order of 10–20 percent of the cost of a comparable conventional home. They understand that if you could build a solid wood house for half the cost of the "stick house" next door, the countryside would be littered with log subdivisions instead of the other way around.

COURTESY GASTINEAU LOG HOMES

Log homes are becoming larger and more sophisticated. Complex rooflines, multiple dormers, and irregular outlines require intricate engineering details and substantial construction skills.

OWNER-CONTRACTING

A professionally built home usually falls into one of two categories. It's either builder-built or contractor-built. Builders employ their own crews, have their own equipment, and range from four guys and a pickup to sprawling office complexes with armies of workers and fleets of trucks. Most builders in the log home construction business are found toward the smaller end of the spectrum.

General contractors, on the other hand, usually do not employ workers directly. They rely instead on a variety of subcontractors that specialize in one aspect of home construction. The "trades," as they're called, include excavating, masonry, plumbing, electric, heating and cooling, carpentry, roofing, drywall, painting, landscaping, and many more. The general contractor (GC) orchestrates the trades. In addition, the GC estimates the cost of the house, oversees the budget, hires and fires subcontractors, schedules the work, and sees that everybody gets paid. The owner's construction contract is with the GC, who is responsible for the outcome of the project.

The distinction between builder and general contractor is important to prospective log home owners. It's generally less expensive to work with a builder than a professional GC. The builder employs workers directly and owns equipment. The builder prices his work at his cost (overhead) plus a (usually) reasonable profit. The GC subcontracts; subcontractors actually employ workers and own equipment. The subcontractor prices his work to the general contractor, based on his overhead and a reasonable profit. The GC then totals up the subcontract prices, adds his own overhead plus a reasonable profit, and charges the log home buyer accordingly. So employing a GC means paying a markup on top of a markup.

Be forewarned, however. Smaller builders often become general contractors by default. The builder may employ carpenters and roofers but subcontract the rest of the trades. To the home buyer, this means that, even though you've employed a builder, you are still paying a general contractor's markup on much of the project.

A log home buyer who becomes a general contractor saves that GC markup (10–20 percent) of the final house cost. There may be a savings of a portion of the builder-constructed house price, also, because total subcontractor overhead may be less than builder overhead and builder profit is eliminated.

So the plus side of acting as your own GC is a significant savings in construction costs. Another benefit that is harder to measure is having more direct control over the final product. By dealing directly with sub-contractors, the home buyer is in a better position to monitor the quality of work being done. Finally, there is the satisfaction of knowing your log home with an intimacy you wouldn't have if you turned overall responsibility over to a third party. And there are the stories, too. If you choose to act as your own general contractor, I guarantee you some fascinating stories to tell your neighbors (although it may take a while for you to see the amusement in some of them).

What about the down side of general contracting? First, the time commitment is significant. Before construction begins, you will need to spend time organizing the project, finding subcontractors, getting bids, estimating the final project cost, and scheduling work. This is on top of the home buyer's chore of securing financing. Plan on a month's worth of weekends and evenings to do justice to the preconstruction phase of your

COURTESY GASTINEAU LOG HOMES

Sloping lots offer opportunity for walkout basements. The lowest level of this chalet-style house incorporates basement garages as well as living space. Basement bedrooms and family rooms can be made more comfortable with natural light from above-grade windows in exposed foundation walls. If a walkout is on your "must-have" list, choose your lot and building style carefully. Have your builder or a surveyor examine your lot to determine whether a reasonable slope exists.

COURTESY GASTINEAU LOG HOMES

The log home dream of many—a pastoral retreat set in quiet countryside. As urban society becomes increasingly complex, more people turn to a dream such as this where walls of solid wood create a haven from the outside world. Achieving the dream, however, requires careful planning, detailed preparation, and true dedication.

project. There may be a few early mornings thrown in, too; because good subcontractors are on job sites all day, the only way to reach them may be with a six A.M. phone call.

A second disadvantage is the amount of responsibility you assume. You will be responsible for the final budget of the house. If you forget a material or labor cost, it's your problem. If there is disagreement between you and your subcontractor as to whether something should be considered part of a bid, it will be your responsibility either to straighten the sub out or come up with the extra cash to complete the task.

Managing subs is not always an easy task. There are subcontractors in any region whose greatest skill is avoiding doing the work they agreed on for the price they agreed to. It takes a combination of patience, assertiveness, and common sense to manage. Be prepared for some amazing excuses for absences and some incredible explanations for shortcuts (especially during hunting season and fishing season, and when the weather is extremely hot, cold, unusually wet or dry, or exceptionally nice; in other words, almost anytime). This is not to say that there are not excellent, responsible people in the building trades — there are. Unfortunately, there are also some accomplished pretenders.

Finally, there are always a few tasks — some unpleasant — that fall by default on the general contractor. For example: There's a sudden turn in the weather over a weekend. It's Sunday morning, and a beautiful spring week has suddenly become a weekend of snow squalls and sleet. The front came in with strong gusty winds, and the plastic sheeting that covers your building materials is probably in shreds. The weather is expected to be unreasonable for the entire coming week. Somebody has to drive the thirty miles of slick roads to your job site to secure the materials and replace the torn tarping. Don't hold your breath waiting for volunteers.

About Log Homes:
Myths and Realities

IN THE COMING PAGES we will cover in great detail what makes a log home. Before launching into that, I think it good to devote some attention to what a log home is not. Perhaps it will prevent potential log home buyers from suffering the misery of embarking on a dream that contains little or no possibility of becoming real.

A log home is not a form of low-cost housing. Put another way: A cheap log home is a cheap home—period. The skills and processes necessary to produce a log home that will please a modern owner are every bit as sophisticated as for any other type of housing. Log homes can require more skill and time to trim. Also, there may be much more stained and finished woodwork in addition to logs, adding further to finishing costs. In my experience, rarely will a log home cost less than a comparable conventional home. Costs generally run about the same to 10–15 percent more, depending on the homeowner's taste and the design of the house.

A log home is not an unchanging, inflexible structure. Wood is a dynamic building material, and large timbers behave very differently from dimensional lumber used in stick-built homes. Rarely will you find the uniformity characteristic of a conventional home. Good log home builders and manufacturers understand their medium, and design and build accordingly. If you expect machine-like quality and workmanship throughout and a house that will be free of maintenance and occasional adjustment or repair, you will be disappointed. Most log home shoppers seek the individuality and the variability characteristic of log homes.

A modern log home is not a primitive structure. In early America, log homes were "starter" houses. Once the land was cleared and the family prospered, it was usually a priority to get out of the log "cabin" and into a "house." Most early cabins were dark, dusty, and drafty; our forebears would think us out of our minds for choosing a log home. It is modern technology, engineering, products, and materials that have transformed the primitive log cabin into a contemporary home of comfort and beauty.

MAINTENANCE

Long ago, log homes were touted as requiring no maintenance. As the number of log homes increased and the industry gained experience, it has become apparent that the no-maintenance log home is a fiction. Realistically, a log home, like any other home, requires some care to maintain its appearance, energy efficiency, and structural integrity. Now we hear about "low maintenance" log homes. No reputable manufacturer will try to sell a log home as maintenance free.

Log home maintenance is mostly preventive, aimed at protecting logs against their natural enemies: water, insects, and decay. Much of this maintenance is similar to that for a conventional home: periodic insect inspection, inspection for decay and water damage, and application of exterior preservatives. A variety of water repellents, preservatives, and stains are available to help logs shed water, resist insects, and maintain their color. To be effective, treatments must be applied regularly. In this respect, a log home requires more attention than a vinyl- or aluminum-sided house, but the maintenance is not difficult and most people consider the compensation of a solid wood home more than enough to cover the little additional time and money involved.

Since most maintenance issues arise during and after the building of the house, Chapter 10 includes further discussion of how to care for your home properly. The maintenance required for your roof, however, is a little different. Most log homes today use the same roof coverings as conventional houses—fiberglas or composition shingles over tarpaper. Some people choose to spend more money for cedar shakes, slate, tin, or tile.

Maintenance will vary with the climate as well as the type of roofing material used. The bottom line on maintenance should be to follow the manufacturer's recommendations.

Modern fiberglas and composition shingles require little maintenance. Once the shingles are in place and have been sealed down by the sun, the only maintenance required is to watch for damage from wind or ice. Simply keeping the roof in good repair will give it maximum life.

Cedar shakes are beautiful to look at but require a bit more care. Unlike conventional shingles, cedar will show the effects of weather. Some people like the gray, weatherbeaten look of an old cedar roof, but others want their shakes to retain the color of a new wood roof. This requires regular treatment of the roof with protective chemicals. The frequency of treatment will vary with the siting of the house.

Slate roofs are extremely durable, but they are costly. In addition, the weight of a slate roof requires consideration when engineering the house and during framing. Talk with people who have slate before deciding. If you do go to the trouble and expense of a slate roof, there should be less maintenance than with a conventional shingle roof.

Tin roofs are becoming more popular. These roofs are long lasting, require little maintenance, and are not as costly as slate, though they are more costly than conventional shingles. The gentle drumming of rain on a tin roof is reason enough for some people to spend the extra cash.

Some log home owners mix tin with conventional shingles. They opt for fiberglas or composition over the living area with a tin porch roof, for example. This is one way to get the romance of tin while retaining the cost advantage of conventional shingles.

Tile roofs are popular mostly in the Southwest and West. They are more suited to drier climates. Tile roofs are expensive and require more skill in installation. You don't climb around on a tile roof unless you like replacing tile. If you live in a region where these are common, or if you prefer the look they offer, try to visit some home owners who already have them.

SHRINKING AND SETTLEMENT

Log homes are structurally unlike conventional homes. Large timbers behave very differently over time than stick-built walls. This is something not always recognized by log home manufacturers. Most are aware of the differences and deal with them in their construction methods. Some, however, still cling to the illusion that once the house is up, it's like any other.

Two major factors to consider in maintaining the structure and appearance of any log home are shrinking and settlement. They are not the same thing. Shrinking involves the gradual release of moisture contained in the log. A living tree contains a great deal of moisture. There is water within the living cells and between them. The moisture content of green wood ranges from 30 percent to 90 percent for the species that are commonly used in log homes. (Moisture content is the ratio of the weight of water contained in a given piece of wood to the weight of the wood when it is completely dry.)

Once cut, a log begins to lose moisture, and moisture loss will continue until the wood is in equilibrium with its environment. This means that a log cut from an eastern forest, milled, and used in a log home located somewhere in the East will wind up with a very different moisture content than the same log used in a house on the high, dry plains of the West. Also, depending on the type of wood and its density, it may take many years for the log to reach equilibrium with its environment.

As wood dries, it shrinks. This leads to a second structural consideration: settlement. Unless prevented, as the logs shrink and become smaller in cross-section, they will settle. In an 8-foot log wall made from green logs this settlement can be upwards of 3 inches (up to 6 inches with some species of wood). Before you panic, understand that drying and settlement are not problems in themselves. (Even stick-built homes, constructed using kiln-dried dimensional lumber, will show some settlement.) Problems appear only when the log manufacturer fails to consider the effects of these factors in the engineering of the house. Most log home manufacturers and all reputable manufacturers acknowledge that shrinking and settlement are serious issues in log home building. How they deal with these issues is a good indication of their experience, sophistication, and integrity.

Recognizing that wood shrinks and shrinking wood settles should simplify dealing with these issues. However, wood is a dynamic material affected by environmental variables — predicting the effects of shrinking and settlement, and arriving at a standard solution are not quite as easy as it might seem. Different species of wood behave differently, and logs from the same species respond differently depending on where and when they were cut, how they were stored, the age and location of the stand of trees from which they originated, and the soil conditions where they were grown. The environment of the log home must be considered, too. A moisture and temperature regime that varies greatly from the one where the logs originated can greatly affect how the logs respond. Engineering and building to accommodate these factors becomes a matter of dealing with a range of potential responses.

Shrinking is greatest immediately after cutting and slows as water is removed. For this reason, many manufacturers dry their logs for a specified period before shipping them for use in home building. Other manufacturers kiln-dry their logs. Both treatments can remove substantial amounts of water, reducing shrinking considerations in the log wall. The important thing to remember is that neither one of these treatments alone eliminates all potential problems. Some people make the assumption that using kiln-dried logs eliminates all concerns over shrinking and settlement. Consider, however, that kiln-dried logs are not dried to the same moisture level as dimensional lumber. The time required to kiln-dry a large timber is substantially greater than that required to dry a load of 2x4s. So even a kiln-dried log may show some signs of shrinking. Also, if the drying procedure takes the moisture content of the wood below the average moisture level of the log home's environment, the log may actually swell. Drying alone, either air or kiln, does not automatically eliminate shrinking and settlement concerns.

Settlement is only partly related to dryness of the logs. The structural system and construction methods used both play an important part in settlement. Most manufacturers accommodate these factors in the design and construction procedures they employ. For example, many manufacturers include a "settlement space" above windows and doors in log walls. The size of this space varies with the type of wood, its moisture condition, and the structural system used. The function of the space is to allow the log wall to settle without affecting windows and doors. In the past, log home manufacturers and builders who did not incorporate settling features have, after several years, faced windows and doors that were difficult to operate. In extreme cases, glass in windows has even cracked. A simple settling space allowed over windows and doors during construction can avoid major headaches and expense later. Settling spaces range from one to six inches, depending on the type of log and structural system. The log home manufacturers will specify the amount for their product. Settling spaces are usually filled with loose insulation and sealed from weather. Inside, they are hidden behind trim. As logs settle, the insulation simply compresses. In dealing with the issue of settlement, remember that while it may be a characteristic of log homes, it isn't really a problem, as long as the manufacturer and builder account for it during design and construction.

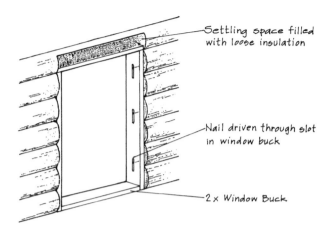

Settling space filled with loose insulation

Nail driven through slot in window buck

2 x Window Buck

DEALING WITH SETTLEMENT: WINDOWS AND DOORS

Some log home manufacturers are beginning to use sophisticated fastening systems in the log wall, ranging from springs (compression packs) that slip over standard galvanized spikes, to threaded rods that run the height of the wall (through-bolts). These systems are designed to maintain the tightness of the log wall. (Almost anyone who has done any serious searching for log homes has come across at least one "I could fit my hand between the logs" experience. Again, species of wood, engineering system, construction methods, and quality all figure in such incidents.) Most fasteners have been in use for only a short time, however. The problems they are designed to address usually show up after a relatively long period of time. Therefore, the jury is still out on their effectiveness. Listen carefully to the sales pitch of a company using such systems, then listen to the rebuttal of someone who doesn't use them. Ask for evidence in the form of a structure standing for at least five to seven years. Finally, consider that there are plenty of log homes that are doing beautifully without such high-tech innovations.

Settlement of a log wall can affect not only the wall itself, but the whole house structure. Most manufactured log homes use conventional framing for interior bearing and non-bearing walls. Conventional framing does not settle to the same degree as some log walls. Unless this is taken into consideration during design and construction, the result can range from a slight hump in the ridge of the house to roof trusses being supported only by interior partitions. In two-story houses, the second floor may gradually develop a slope. In a 24-foot-wide house, for example, with log walls that settle 2 inches over a ten-year period, the second floor can develop a slope of 2 inches over 12 feet, or slightly less than ¼ inch per foot. A 3-foot doorway parallel to the direction of the slope will gradually become as much as ¾ inch out of square. This means reframing the door opening or planing the door to a noticeably irregular shape. A wider house would show less slope, and a narrower house would show more.

Some log companies have incorporated features that should eliminate even minor unevenness between inside and outside points. Usually these take the form of adjustable jacks or shims that can be removed from interior supports as exterior walls settle. These systems require that interior partitions not be fastened securely to ceiling framing but allowed to float free. Partitions are usually held down but allowed to float free, leaving a settling space somewhere between floor and ceiling. The space between partition and ceiling is concealed behind trim that slides over the surface of the partition as settling occurs. In these systems, maintenance consists of periodically checking the amount of settlement and removing shims, adjusting jacks, and tightening through-bolts, as called for by the manufacturer.

Before you assume that a warranty will provide an easy out in case your logs don't behave as you would like, beware. Because these are natural characteristics that can occur despite the manufacturer's and builder's best efforts, they are excluded from most warranties as "acts of nature." Basing your log home decision on a warranty is false economy. Instead, rely on research — visiting log homes and talking with owners and builders — your own common sense, and the reputation of the people you are dealing with. Consider that when someone tells you shrinking and settlement are not concerns for his or her product, you have no way of knowing, short of extensive research, whether that is true. On the other hand, when someone acknowledges these factors and points out the engineering and construction methods used to accommodate them, you can be assured that you are dealing with someone concerned with the quality of the product. No manufacturer, sales representative, or builder is going to expend the effort necessary to deal with a potentially negative sales point without good reason.

In summary, shrinking and settlement of logs are issues that need to be considered in the design and construction of log homes. However, the number of systems and variables makes it impossible to say which system

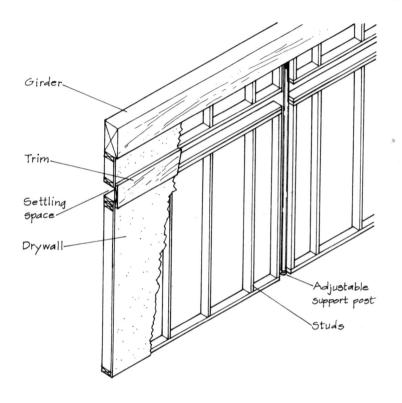

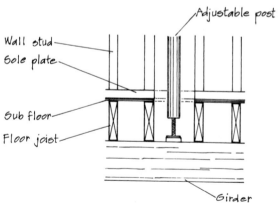

DEALING WITH SETTLEMENT: INTERIOR PARTITIONS

is best. Listen to sales representatives and read the companies' sales literature. Remember, it's not simply a matter of engineering, it's also a matter
of what you will be satisfied with. As a general rule, the smaller and
simpler its design, the less a house will be influenced by shrinking and
settlement.

ENERGY EFFICIENCY

During the energy/environmental crisis of the mid-1970s, high energy
costs and a heightened awareness of environmental degradation pushed
many people to seek alternative lifestyles. A self-sufficiency ethic developed that had many Americans turning "back to the land"—growing and
preserving their own produce, raising their own meat, and building their
own homes. Nothing fit the image of a simpler life quite like a sturdy log
cabin in the woods.

In addition to its natural appearance, a log cabin had the appeal of
do-it-yourself construction and energy efficiency. A log home in a remote
clearing, snug against the weather and heated by an efficient woodstove,
became the dream of multitudes. For those of the multitudes who weren't
quite ready to abandon forced-air heat and give up Saturday golf for the
rigors of cutting wood and canning vegetables, a weekend log retreat
offered the perfect compromise—simplicity on a part-time basis, simplicity without the down side.

A mystique has grown up around the superior energy efficiency of log
homes. Many people relate stories of heating their entire log home on
several cords of wood, resorting to back-up fuel furnaces or baseboard
electric heaters only during periods of extreme cold. A large chunk of this
mystique is based on misinformation and misinterpretation, however. Log
homes *can* be more energy efficient than conventional housing, but not for
the reasons that many people think.

Wood is not a particularly good insulator. It is poor enough that
during those early energy-crisis days, log homes could not obtain government-backed financing because the R-values of log walls did not meet
requirements for energy efficiency. A 6-inch-thick white pine log wall has
an R-value of approximately 7.9; a western red cedar log boasts only
R-6.5. A standard 2 x 6 framed wall system, on the other hand, can be
insulated to R-19 + (depending on the siding, sheathing, and interior wall
covering used). On the basis of insulating ability, as measured by
R-values, log walls simply don't measure up.

What does this mean for log home shoppers? It means that basing a
choice of wood species on R-value is more than a little misguided. If sales

people tout their logs as having a higher R-value than those of their competitors, ask if they offer a 16-inch-thick white cedar log, because, unless they do, they are not offering anything equal to 2 x 6 framewall construction (as far as insulating ability goes). So where do these claims of superior energy efficiency originate? As it turns out, in the matter of energy efficiency, log homes are a little like bumblebees.

Bumblebees, according to the aerodynamic engineering calculations upon which the flight of airplanes is based, can't fly. Plug a bumblebee's vital statistics into the calculations, and it's strictly a hiker. According to the numbers, it does aerial acrobatics only in its dreams, which means you must be imagining it up there in your rafters.

So it is with log homes. While log homes have low ratings according to standard insulating calculations, they have been proven the equal of conventional insulated framed wall systems. As the modern log home industry built up a track record, it became apparent that log walls were equaling and even outperforming conventional wall systems, despite the statistics. In order to be included in government-backed financing programs, the log home industry mounted an effort to demonstrate energy efficiency. They lobbied for government testing until finally the National Bureau of Standards undertook a long-term comparison of log walls with other accepted types of home construction. When the results were tabulated, the log home industry had its proof. A 6-inch log wall equaled or exceeded the energy performance of any other type of wall during all seasons tested except the dead of winter (here an insulated framed wall won by a small margin).

The reason for the successful performance of logs was determined to be "thermal mass"; that is, the mass of the logs allowed them to absorb, store, and release heat over a period of time. A log house is actually a form of passive solar home. This is an important fact to keep in mind while designing a log home for energy efficiency. Too many people see the tag "energy-efficient" as a license to ignore many of the building limitations necessary to benefit from the inherent energy-saving ability of a log home. They run amuck with energy-inefficient cathedral ceilings, glass in the wrong quantity and the wrong place, and disregard for the role of house siting in maintaining energy efficiency. Someone who builds a log house with a north-facing, heavily glassed, cathedraled living room is not going to have an energy-efficient home.

To obtain maximum benefit from the energy-conserving character of log homes (and reduce construction costs): First, keep the house as small as practical. Large homes and large rooms are expensive to build, furnish, clean (in time if not money), and heat and cool. You'll give up most of the cozy, enveloping warmth that only a log home can offer if you insist on the house and room sizes that the home magazines generally tout. Consider how each room will be used and size it according to the furnishings necessary to accomplish that function.

Second, keep cathedral ceilings to a minimum or eliminate them entirely. Cathedrals are beautiful to look at, but unless you can fly or spend a lot of time sitting atop a ladder, the energy to heat and cool those spaces is largely wasted. Ceiling fans may offer a partial remedy, but they will not recapture all of the expense. If you must have cathedrals, locate them carefully.

Many people mix cathedral ceilings with large expanses of glass. If this is your desire, site the house carefully, facing the glass and cathedral area as close to south as possible. The further away from south that your glass faces, the more dollars you are sacrificing for appearance's sake.

Third, keep the house design simple, with few angles and corners and a simple roof line. The amount of surface area where energy loss occurs (walls and roof) will have a great effect on your heating and cooling bills. Consider three log houses, each with 1,500 square feet of living area. One house is a simple one-story rectangle measuring 30 x 50, with a low pitched roof and flat ceilings inside. A second house has two stories with no cathedral ceilings. It measures 25 x 30 and has the same roof pitch as the first house. The third house could be called a 1½-story house. It measures 25 x 42.8 with a steep pitched roof. It contains a cathedral ceiling along 20 feet of its length. The remainder (22.8 feet) of its length contains a second floor. To obtain enough living area with 8-foot ceilings on the second level, a dormer has been added to the back of the house.

How will these houses compare in terms of energy use? Consider that the principal source of energy loss will be through the roof. The example assumes the same amount of glass area for each house, with all houses on the same type of foundation. Here is how they compare:

	One-story	Two-story	1½-story
Square feet of living space	1,500	1,500	1,500
Linear feet of exterior wall	160	220	138
Cubic feet of conditioned space	12,000	12,000	15,122
Square feet of roof area	1,581	791	1,352

Notice that the two houses without cathedral ceilings have the same volume of conditioned space. The house with the cathedral ceiling has an additional 3,122 cubic feet of space to heat and cool, with no increase in usable living space—more than 25 percent greater heating and cooling requirements.

Next, compare roofs, the main source of energy loss. The two-story house without a cathedral has half the surface area of the single-story home to absorb heat in summer and lose heat in winter. It has 561 square feet or 41 percent less roof area than the 1½-story house. This should give some idea of the premium paid for cathedral ceilings.

These comparisons are good not only from an energy-efficiency standpoint, but also from a construction-cost standpoint. Consider how the houses compare in amount of required roof framing, foundation, excavating, and so on.

There are other factors to consider in maximizing energy efficiency. For example, fireplaces are beautiful to look at but tremendous sources of energy loss. The most efficient fireplace can't begin to compare with most woodstoves for energy efficiency. The inefficient burning of wood in fireplaces is leading in some areas to restrictions on their use. Expect to see further curtailment in use of fireplaces near urban areas in the not-too-distant future.

If a fireplace is a must, consider a "zero-clearance" unit, a manufactured firebox, often with air-circulating capability, that can be enclosed in wood framing. The framing can be sheathed and covered with artificial stone or wood. These are the types of fireplaces that appear in most modern homes. In addition to being more energy efficient than masonry fireplaces, they are generally less expensive.

If you want to use real stone for your fireplace and chimney, consider moving it inside the house. Stone has great thermal mass and conducts heat slowly. During winter, a stone chimney built into an outside wall will be continually drawing heat from the conditioned space inside. Instead, put the stonework inside the living area, where it is surrounded on all sides by conditioned air. Better still, locate it to receive direct sunlight on winter days. This will allow you to use the thermal mass of the stone to help redeem some of the dollars lost in choosing a fireplace over a woodstove.

Roof overhangs contribute to energy efficiency in log homes. A wide overhang (2 feet or more) blocks the direct rays of the high summer sun, while allowing passage of the direct sunlight from the low winter sun. In addition to aiding your energy budget, wide eaves will help prevent carpet and curtain fading from the sun's ultraviolet rays.

A number of log home companies now offer "super-log" systems. These systems take the thermal mass advantage of a log wall and combine it with the insulating abilities of a frame wall. At their most basic, these systems consist of a 2 x 6 conventional insulated frame wall sandwiched between half-logs on the outside and inside. Other systems use logs that have been hollowed and the cavities filled with insulation. Some of these walls boast energy performance above either solid log or heavily insulated conventional framing.

Your choice of a heating and cooling system will affect your monthly energy costs. In my area, many people select electric heat pumps for an economical compromise in installation cost and monthly energy bills. Heat pumps work on the principle of heat exchange. By circulating heated air against cooler air, some heat is transferred or exchanged. If the range of the heat pump is exceeded, an auxiliary electric or gas heater kicks in to assist. The main complaint of most people is that the air coming out of the heating ducts is usually not warm (it's about room temperature). This means giving up the little luxury of warming your bare feet over a heat register on a wintery morning.

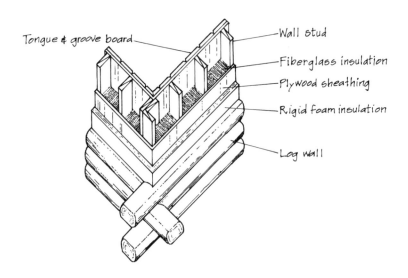

Tongue & groove board — Wall stud — Fiberglass insulation — Plywood sheathing — Rigid foam insulation — Log wall

INSULATED LOG SYSTEM

A forced-air, electric, gas, wood, fuel oil, or coal furnace presents another alternative. These furnaces also rely on systems of ducts to deliver heated air. Depending on the region and the current energy-resource situation, these systems may or may not offer bargains. It's best to discuss your needs and house plans with an energy consultant at your local utility company or government energy office and collect sales literature about the types of systems that interest you. Talk to people who use similar systems and compare costs.

You may also want to consider one of the radiant heat systems. Radiant systems do not rely on ductwork to carry heated air; they radiate heat directly. Pipes carrying hot water beneath floors and through baseboards warm air directly. These systems are just beginning to make their way into the log home world, and their merits have yet to be decided. Research and an open mind may lead you to a system that saves money and suits your lifestyle. Make this an important part of your log home planning process.

Construction methods and quality of construction rank almost as high as proper design in determining the energy efficiency of your log home. A sound energy-saving design is useless if the builder doesn't use appropriate construction techniques. Don't let anyone tell you that putting up a log home is no different from erecting a conventional house. Logs present a much more dynamic building system than framing. Logs move, shrink, and settle. A builder experienced in log home construction is much more likely to understand the manufacturer's sealing and settlement system than someone with only conventional framing experience. Even if the manufacturer provides a thorough set of construction details, situations will arise in which a familiarity with the behavior of a log wall system can mean the difference between a tight, energy-efficient log home and an energy (and bank account) guzzler. This is especially true of larger, more complex homes.

CAULKING

Another area of controversy in log maintenance concerns the use of caulk. Because even good caulk decomposes or becomes brittle over time, causing breaks or separations that must be patched, purists say that the weathertightness of a log system should not depend on caulk. They believe that the logs should be assembled so that tightness is maintained by the quality of construction alone. Good handcrafters may even be able to pull this off.

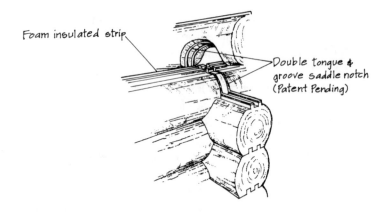

Foam insulated strip

Double tongue &
groove saddle notch
(Patent Pending)

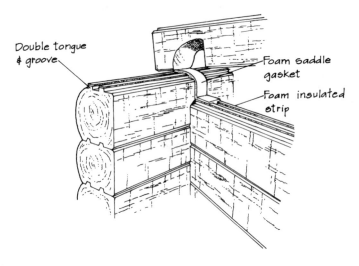

Double tongue
& groove

Foam saddle
gasket

Foam insulated
strip

SEALING A LOG CORNER

(COURTESY HERITAGE LOG HOMES)

However, we are not dealing with handcrafted log homes here. Because of the methods used in manufacture, variable site conditions, and variable construction quality, anyone who expects to maintain the weathertightness of a log home without the use of caulks and modern sealants has been misinformed. A modern log home is a twentieth-century product. Its energy efficiency depends as much on twentieth-century tools and technology as it does on the mass of the log walls. Follow your manufacturer's construction guidelines on the use of caulks and sealants. Once the home is finished, inspect the seals periodically and touch them up as needed.

Log Home Kits

THE FIRST CHOICE most log home shoppers face is what type of log home they want. Today's log home marketplace can be divided into two broad categories: handcrafters and manufacturers. The names are fairly self-explanatory. Handcrafted log homes are individually built, usually by a person skilled in taking raw logs or even standing trees and turning them into a home. Handcrafters often rely on tools and techniques that have existed for centuries, combining them with modern tools and engineering knowledge to produce one-of-a-kind homes. Handcrafted log homes can be objects of singular beauty. Their chief drawback is the expense and limited availability of experienced craftspeople. I will not dwell on hand-crafted homes in this book. Publications listed in Chapter 11 can direct you to handcrafters and information about handcrafted homes.

One of the main factors enabling log homes to grow from tiny cabins to sprawling residences has been the growth of the log home industry in which manufacturers cut, shape, and mill raw logs and package them as "kits." As they include milled logs and construction components, log home kits offer a way for people with limited construction experience to under-take projects beyond their abilities and time constraints. The quality control offered by log home manufacturers improves the chances of getting a high-quality, energy-efficient home. The service of offering material packages, including engineered blueprints, eliminates two major obstacles facing would-be owner-contractors—estimating material requirements and figuring out construction methods.

Kits, or packages, range from truckloads of logs and fasteners to complete structural systems including doors, windows, roofing materials, and more. Kits offer several advantages to prospective log home buyers, such as lower costs and generally easier construction methods. Using milled logs and standard components, generally allows construction to proceed faster (and therefore cost less). Manufactured kits often produce a more finished, somewhat less rustic look than handcrafted homes. Some buyers find this appealing; others do not.

Manufactured kits can be divided several ways: by shape or type of log, by completeness, and by whether logs are precut. Logs may be peeled or milled. Peeled logs are essentially trees that have had the bark and outer sapwood removed. Round, hand-peeled logs have a faceted surface characteristic of log homes of the western frontier. Large, squared timbers may bear the adz or axe marks (real or simulated) characteristic of eastern pioneer dwellings. Some types of corner joinery do not allow logs to rest tightly against one another. Instead, a space is left that is filled with weatherproof "chinking," the modern equivalent of the mud and straw used to seal frontier dwellings.

Milled logs are run through a sophisticated planing machine that removes the outer sapwood and imparts a variety of shapes. Milled logs may be flat or rounded on one or both sides; they may have outer surfaces beveled to resemble siding. Wood grain will be more noticeable on milled logs, and surfaces will be smoother and more uniform. Most milled logs include some feature for interlocking with adjacent logs—single or double tongues that fit into matching grooves on adjoining logs, or splines (narrow strips of wood, plastic, or composition material) that fit in slots on adjoining logs to prevent air and water filtration.

SHOPPING FOR A KIT

There are a multitude of log home companies in the United States and Canada, each striving to put *you* into their product. A multitude of sales literature touts the superiority of various wood species, corner systems, sealant systems, and structural systems. It's easy to be overwhelmed by the variety of log home kits available. Having studied the industry, produced advertising for log home manufacturers, and built a number of log homes, I've come to several conclusions.

First, virtually any major manufacturer that has remained in business for any length of time produces a reputable product. Its homes *can*, if constructed properly, live up to the stated claims. The success of a log home depends as much on the skills and experience of the builder as it does on the manufacturer.

Second, most log home shoppers begin the search for their dream home with the sales literature of the manufacturer. For too many, that's where the research and planning end. Keep in mind, as you begin working to make your log home dream a reality, that you will be living in a *home*, not a *kit*. A log home package is a bunch of materials only. Even if the company offers an erected shell, you are still a long way from a completed home. There are many log home shoppers out there who spend 80 percent of their energy choosing a log company and devote a rushed 20 percent to getting the house built and finished. This is exactly backwards.

Over the years, I've met a number of log home enthusiasts who have spent years researching various companies, even traveling tens of thousands of miles to visit hundreds of manufacturers. They usually have definite opinions about which wood, structural system, corner style, and fastening system is superior. But I have yet to meet two with identical opinions.

Here's how I would approach shopping for a kit manufacturer. First, I would choose the appearance or "look" that I liked—whether it's huge squared timbers, round hand-peeled logs, or milled logs. Milled logs come in a variety of shapes including rounded or flat on both sides, beveled on the exterior to give an appearance similar to wood siding, and the popular "D" log that is rounded on the outside surface and flat on the interior.

With my log style chosen, I would locate and visit several homes representing different manufacturers. I would consider the appearance, type of trim (if it's part of the kit), and components included with the package. I would visit with the sales representatives and give much consideration to "gut" reactions to the various people I speak with. Do they seem straightforward, interested, courteous, able to make suggestions, or are they high-pressure, hard-nosed, and full of claims of superiority and economy that seem a little stretched? There's nothing wrong with believing your product is best (mine is actually best, incidentally), but good sales reps recognize when someone is looking for something that they can't offer and say so.

Finally, with my selection narrowed to one or two manufacturers, I would approach several log home builders and obtain estimates for a

Flat on Flat

With Spline

Rectangular

CORNER SUPPORT SYSTEM WITH CHINKING

Round on Round
with chinking

With Spline

Round

TONGUE & GROOVE

Single Tongue & Groove

Double Tongue & Groove

Ripple (or Triple Tongue & Groove)

Concave over Round
(Swedish Cope)

HORIZONTAL INTERFACE SURFACE STYLES

(COURTESY LOG HOME LIVING)

ROUND LOGS

Full Round

Swedish Cope

'D' LOGS

Flat & Round

Round & Flat

Round & Round

SQUARE LOGS

RECTANGULAR LOGS

Normal

Ship Lap

Normal

Bevel Edge

LOG PROFILE STYLES

(COURTESY LOG HOME LIVING MAGAZINE)

completed house. I would, under no circumstances, base my final decision on where I am going to spend my days, on the price of the log kit only. It would be like choosing a conventional home based on which one came from the cheapest lumberyard. With that said, let's examine some of the major selling features of log home packages. This is background information to help you evaluate your desires.

Wood Species

One of the questions many log home shoppers ask is, "Which is the best wood for a log home?" At a log home building seminar several years ago, an attendee actually became angry and abusive because the seminar leader would not name *the* best wood species. But I agree completely with the leader that there is no "best" wood. Every wood has properties that are advantageous and properties that are drawbacks. Any wood currently used in a properly constructed and maintained log home will outlast you and several generations of heirs. If the home is not properly built and maintained, no wood will last a single lifetime.

Here is a brief description of woods used in log homes:

Cypress
Light yellow to brown heartwood; heartwood of old growth is extremely decay resistant, new growth somewhat less; moderately strong, expensive.

White pine
Eastern variety is light in color, straight-grained, easy to work. Its stability makes it a popular wood for many uses. Difficult to find in long straight lengths, resists twisting.

Yellow pine
Heartwood is reddish brown, easy to work, moderately decay resistant, more prone to shrink, twist, and warp than other species, relatively inexpensive.

Red cedar
Reddish heartwood, lightweight, soft, low strength, relatively expensive.

White cedar
Light-colored, decay-resistant, soft, low strength, relatively expensive. Limited to small diameter logs.

Lodgepole pine
Light yellow to nearly white, low strength and shock resistance, moderately priced, available in longer straight logs than many other species.

Spruce
Nearly white, straight-grained, light in weight, low strength and shock resistance, often knotty.

Douglas fir
Very strong, available in larger sizes, moderately priced; often used to make heavy timber trusses and for beams to span long distances.

Oak
Beautiful grain, extremely strong, available mostly as milled logs from few suppliers; price comparable to most other woods. White oak is highly decay resistant.

Hemlock
Eastern hemlock has pale brown heartwood, is moderately heavy and hard, has less tendency to twist, bow, or check than many woods.

Corner Systems
There are a number of styles of corner joinery used in constructing log homes. Each has claims of superiority, and each has drawbacks. As with consideration of wood species, the nature of a log home is such that any corner system in use by a reputable manufacturer will stand well beyond the time you will be concerned about it. I would suggest choosing a corner system based mostly on the appearance that you like. Then invest your energy in finding a builder who will build it right.

Here are brief descriptions of some of the more basic corners used:

Butt-and-pass
As its name implies, one log butts against the side of another log that passes through to overhang the corner on the outside of the house. "Butt" logs alternate from one wall of the corner to the other in consecutive courses (rows) to produce a characteristic crisscross look on the house exterior corners. Its main advantage is ease of construction; its main drawbacks are that it requires strict attention to detail to seal it properly

and corner logs must be fastened very securely against the possibility of twisting.

There are many variations of butt-and-pass corners, often involving a notch or V groove cut into the pass log to receive a tongue cut into the end of the butt log. Such embellishment may increase the weathertightness of the corner as well as provide additional resistance to twisting.

Notched

Notched corners have notches cut in one or both logs of the corner. Depending on the style of corner, the result may be corners with solid overhangs in both directions of the corner or crisscrossing similar to butt-and-pass systems. Notched corners are stronger and hold corner logs better than butt-and-pass. They may be easier to seal, providing the notches are well made. Notched corners may involve some additional construction time, depending on their engineering.

Post

Post corners use a corner post that is slotted to accept tongues cut into the end of corner logs. This type of corner has no overhangs and requires different engineering treatment from other log corners, because the post will not settle with the logs.

Dovetailed corners

Dovetailed corners produce a distinctive appearance. The ends of logs to be joined are cut into wedges with corresponding angles to produce an interlocking corner with no overhang. Because of the angular nature of the joinery, it is almost always used with square or rectangular logs. Dovetailed corners are strong. As with all corner systems, care must be taken to obtain a weathertight seal.

Fasteners

Log homes are held together by a variety of fasteners. Spikes are probably most commonly used, followed by lag screws. These fasteners may be supported by other devices aimed at controlling or altering settlement characteristics of the wall. Compression packs are springs placed beneath the head of spikes to exert a steady pressure on the logs as they settle. Through-bolts are threaded rods that pass through the entire log wall from top to bottom. They are designed to be tightened as the log wall settles, to help maintain the integrity of weather seals and log joints.

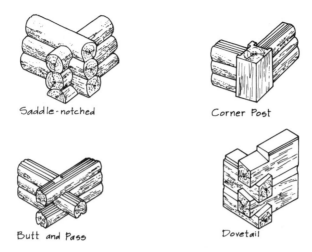

Saddle-notched

Corner Post

Butt and Pass

Dovetail

CORNER SYSTEMS

(COURTESY LOG HOME LIVING MAGAZINE)

Sealants

Achieving and maintaining a good weather seal depends on more than log joinery and fasteners. Virtually all log homes rely on some sort of sealants to eliminate air and water infiltration. Sealants include caulks, expanding foams, foam gaskets, and splines. These are used individually or in combination. A review of log home sales literature over several years quickly shows that sealant technology and methodology are not set in stone. As the log home industry ages, it is capitalizing on new technology and learning better ways to do things to produce homes of higher quality. If you happen to visit a log home more than a few years old and hear comments about draftiness or unexpectedly poor energy performance, don't immediately discredit the manufacturer. Consider the builder and the state of the sealant system. Research what the company used then compared to what it uses now.

COMPARING PRICE QUOTES

Perhaps the most frustrating challenge facing log home shoppers is comparing price quotes from different companies. It seems that no two kits are alike. Each company has its own idea about what should be included in a log home package, so that one manufacturer's standard item is another's option and a third manufacturer may not carry the item at all.

Complicating the whole process is the fact that log home manufacturers do not all use the same building system. What may be necessary to construct one manufacturer's kit may not be required to complete another's. I've had customers call me pleading, for example, "Help! What is a fly rafter? Do I really need six? How much do they cost? Will it be cheaper if I get them locally?" Here, then, is a method for beating a path through the thicket of kit comparisons without getting scratched.

Manufacturers follow several different philosophies in arriving at their materials lists. Some include only the logs and heavy timber items such as beams and necessary fasteners, assuming that conventional building materials such as shingles can be purchased locally. Manufacturers using this philosophy generally have very attractive kit prices. But many components must be added to the kit to complete the house structure. By the time everything is accounted for, the final cost may equal or exceed that of a more complete kit.

Other manufacturers offer a "dried-in shell kit." These are more complete but should be evaluated carefully. Manufacturers have differing ideas about what constitutes a dry-in. Some include roof systems right up to and including shingles. Others stop at the tar paper. Windows and exterior doors are part of a dried-in house, but what about the trim? Some kits will include it, and some won't.

Finally, some manufacturers try to follow the one-stop shopping approach and include everything necessary to complete the structure of the house. These packages often carry the outward appearance of being the most expensive. However, by the time all of the additional items necessary to complete one of the less expensive kits are totaled, the cost may exceed that of the seemingly more expensive kit.

Most log home shoppers are not experienced builders, so they often lack the detailed knowledge necessary to evaluate the completeness of a kit. It's easy to get lulled into a false sense of security by impressive diagrams or materials lists provided by a manufacturer. If you're not careful, you may overlook the primary question that you must answer to obtain an accurate price comparison: "What will I need that's not on this list?" Only by filling in the missing blanks for each kit you are considering can you discover which package offers the best value.

The secret to obtaining a realistic comparison of log home packages is to follow a systematic approach. First, decide what actually needs comparing. If your objective is a turnkey log house (one that will be completely built and finished by a professional builder), a comparison of kit prices may be largely irrelevant. What the turnkey home shopper needs to compare is the bottom line finished house price. This is obtained by taking the kit prices with blueprints and materials lists to the builders you are considering. They will provide a turnkey quote that includes all work that you specify. All you need to do is compare turnkey bids.

If you are planning to erect your own log package or act as your own general contractor, price comparisons are more difficult. First, make sure your price quotations each have a detailed material list attached. You must know exactly what you are purchasing. A good book on carpentry (not just log home building) is invaluable here, because it can illustrate and explain terminology in the lists that may be unfamiliar to you (see Chapter 11).

Break the materials lists down into sections. Because manufacturers do not organize their materials the same way, it may be easier to establish your own section headings. Section headings that I use are: 1st Floor System, Wall System, 2nd Floor System, Roof System, Interior Framing System, Dormers, Porches and Breezeways, Decks, and Garages. Beneath each section, list the items necessary to complete that system. For example, under 1st Floor System, I list the following: first floor support posts, first floor girder beam, sill sealer insulation, treated sill lumber, band lumber (to surround the subfloor), floor joists or beams, joist hangers (necessary in certain situations), subflooring, finished floor covering (wood, carpet, vinyl, etc.). Don't forget to list glues, caulks, nails, screws, and other fasteners. They are a necessary part of the building system, too.

Then, use the Kit Comparison Sheet in the Appendices or set up a chart with sections and individual items in a column and manufacturers you wish to compare in a row across the top. Next to each item in the comparison sheet, place a check if the manufacturer includes it in their kit. If it is not included, determine whether the item is needed or desirable (by referring to a carpentry text or asking your sales representative or a carpenter to describe the item's appearance and function). If you feel the item is necessary, though not included, call a lumberyard or supplier to get a price quote for that item. Place that price figure in the appropriate place in the comparison sheet. You can omit from the comparison items that are beyond the scope of any of the log kits (plumbing or light fixtures, for example).

Part of the difficulty in comparing kits is that one manufacturer's building system may require a specific item, while another's may not. Ask sales representatives to explain any item that is unfamiliar to you. Pay attention not only to what is included, but in what form. For example, two manufacturers may both include interior doors. However, one manufacturer supplies door slabs, wood stock to make door jambs and casing, and door hinges. The other manufacturer includes prehung interior doors. In the first instance, you or your carpenters must make door jambs from scratch and install hardware, a time-consuming process that also requires advanced carpentry skills. In the second instance, doors are already assembled and hung in a premade jamb. Installers need only slip the unit into its opening, align it, and fasten it, a much faster process and within the skills of less experienced people.

A comparison that faces many log home buyers is between different manufacturers' roof systems. Company A, for example, offers a complete timber-framed roof using log rafters and purlins. Its kit also includes tongue-and-groove pine decking to cover the timbers, styrofoam insulation panels with a top nailbase, tar paper, and twenty-five-year warranty shingles. Company B offers only the heavy timber components of a similar roof system. Company C offers a conventional roof system of dimensional rafters, plywood sheathing, tar paper, and shingles. How do you compare the real kit costs? And which one is best for you?

The pricing in this example may be misleading. Let's say package A is most expensive, package B is slightly less, and package C is substantially less than both. To make your decision, consider not only your budget but also the log home style you are after. Before you start comparing prices, consider the look you want your home to have. If you aren't interested in a particular style of house, there's no sense wasting time and energy including it in your comparison.

If a log rafter and purlin roof is something you simply must have, then your comparison will be between kits A and B (unless Company C offers that roof system as an option). In this case, roof system A is complete; roof system B needs decking, insulation, a nailbase for the shingles, tar paper, and the shingles themselves. Price these items locally and from Company B, if they are offered as options. Then, all other things being equal, you have a realistic comparison of kits A and B.

If beams and purlins aren't essential and your objective is simply wooden ceilings, then broaden your comparison to include kit C. You'll need insulation and wood tongue-and-groove siding to cover the underside of the rafters. When complete, house C will have a solid wood ceiling without any exposed heavy timbers.

Your comparison should also reflect the labor costs involved in putting up the respective roof systems. Roof systems A and B require special expertise and will need to be completed as part of erecting the structure of your house. Electrical wiring in the ceilings must be done as the roof is being framed. This requires precise coordination with the electrician and perhaps an electrical inspector. You'll also need to account for any concealed electrical wiring at the time the shell is constructed. These factors may make the cost of those roof systems greater than the conventional roof. In kit C, the insulation and tongue-and-groove ceiling covering need not be completed as part of the dry-in. Also, a conventional rafter roof is within the capability of any good carpenter. If you will be installing ceiling insulation and tongue-and-groove, this may be the least expensive roof system. If these items are to be contracted, you may want to do a more detailed labor cost comparison.

Your comparison should include the grade and quality of items in each kit. Often a more expensive kit actually is a better value because it contains higher-grade materials. For example, one kit may have a top-line window, while another includes an inexpensive unit. One company may base its quote on an 8-inch-thick log against the competitor's 6-inch-thick log. Some companies offer different grades of log or trim.

Some areas I've found that vary widely among manufacturers' kits include: log dimensions (a tall, thin log requires less linear footage to complete a wall than a shorter, thicker log); log gables (many companies include solid log gables only as an option); window and door units (range from cheap to top quality); subfloor systems and dimensional lumber (these may be standard in some kits while part of an option package in others); exterior and interior door and window trim; interior framing; interior doors; stairs; and stair and loft railings. Interior partition wall coverings are rarely included in a basic kit price, and soffit and fascia, shingles, exterior wood preservatives, porch and deck materials are often part of option packages.

Don't assume that dimensional lumber and items such as shingles purchased locally are more economical because of a savings in freight costs. Most log packages are shipped by the truckload. It doesn't cost any more to ship a truck that contains a half-load of logs and a half-load of other building materials than it does to ship that half-load of logs alone. Because the buying power of the log company is generally better than an individual's buying power (log companies often deal directly with manufacturers and sawmills), it may actually be cheaper to purchase even dimensional lumber such as 2 x 4s from the log company.

Avoid being wooed by statements like "free freight" or "free delivery." There isn't a trucking company in the country that will haul a tractor-trailer load of logs free of charge. Log companies offering "free freight" are simply building the freight cost into their kit price. It may work to your advantage or it may not. If the company offers "free freight up to seven hundred miles from the plant," and you live three hundred miles from the plant, you may be paying for four hundred miles of additional freight. *Keep your eye on your bottom line.* A $30,000 kit with an additional freight charge of $3,000 means the same thing to your checkbook as a $33,000 kit with free freight.

Another factor to weigh in any kit comparison is the log home company itself and its sales representative(s). It may be that one kit is slightly more expensive than another but you feel more comfortable with that company. How much is that peace of mind worth to you? One company may include on-site assistance, while another provides a detailed construction manual or video. Which is more important? And how do you include those factors in your cost analysis? Answers to these questions depend on your personal situation.

Finally, consider your building circumstances. Perhaps you are hiring a builder or taking your vacation to erect your log kit. It is probably to your advantage to have all of your materials on site at one time, so a more complete kit may be desirable. On the other hand, you may be erecting your log home yourself on weekends, working when time permits, and paying cash as you go to avoid a large loan commitment. In this case, it may be better to purchase a more limited kit. This avoids not only the larger cash outlay but also eliminates costs and concerns for material storage and protection. Why pay in advance for materials that, once purchased, would sit on the building site for several months before they were needed?

Consider your wants, your needs, and your circumstances carefully. Research thoroughly and keep your eye firmly on that bottom line. You'll be more confident in your selection of a log package. You'll also gain a better understanding of what will be involved in putting up your log home, and you'll have the peace of mind that comes from having a solid grip on this important part of your overall house-building budget.

ABOUT ARCHITECTS

One of the simplest ways to approach the task of designing your log home is to hire a professional — an architect. Many log home companies offer architectural services as part of their package price. However, there are a number of companies that rely primarily on draftsmen. These may be guided by an engineer or architect to make sure the structure of your home is sound.

If you are buying a log home package based on a company plan or a close modification, your architectural requirements have probably been met; likewise, if you are purchasing a design from an architectural plan service. If you are starting from scratch, seeking an individual style, and planning to make a considerable investment in your log home, then you may want to consider an architect.

Recognize that an outside architect means an additional budget expense. Your log home company might be willing to offer a price break on your package if you come to them with a complete set of blueprints, but don't count on it. There are unique aspects to most log home companies' structural systems, so even with a complete set of blueprints, the drafting department will have to redraw the plans to conform to the company's unique structural system.

Before hiring an outside architect, consult the log home company or companies you are considering. Ask about the education and experience of the design staff. They may have an architect or a trained interior designer on staff. Discuss your wants and needs with that person. Ask to see work samples, check references, and consider whether your personalities and tastes are similar. Don't expect an individual with a flair for rustic lodges necessarily to be proficient with a contemporary style. Remember, however, that if you can use the services within the log home company, you are money ahead.

Architects generally base their services on the square footage of the house and/or a percentage of the project budget. Prices will vary depending on whether you want only a set of blueprints or you want the architect to oversee the project to completion. Discuss your needs with and obtain estimates from several architects before making any commitments. Find out particularly whether the architects have any experience with log construction. Show them the construction methods used by the log home company you are considering. Study examples of their work.

There are a couple of advantages to working with an architect in designing a large or elaborate log home. First of all, there's availability. Working with a log home company design department generally limits you to communicating through the mail or by fax. You can sit down with a local architect, take him or her to your building site, and watch the ideas being sketched as you discuss them.

A second advantage is that architects generally provide more detail than in-house design departments. Blueprints come with cross-sections and trim details that might not be found in company drawings. Local architects can point out architectural details such as stairs and railings, corners and beam-wall intersections, and discuss how they should be trimmed. They can sketch various alternatives and let you choose. A log home offers a wider variety of trim options than a conventional home. In conventional homes, trim usually refers mainly to doors and windows. In log homes, there are also settling spaces; more elaborate, often custom-made, stairs; loft railings; intersections between logs and other wall and ceiling coverings; and beam attachment points. In addition, trim in a conventional home is usually painted to blend with the interior surfaces. In a log home, trim is often stained, making it contrast with non-wood surfaces. Because of the possible variations, log home companies may avoid including trim details with their blueprints.

However, not everyone is willing or able to spend the kind of money required for the details an architect can provide, and it is not usually necessary. Most trim details can be worked out on the spot between the carpenters and the home owner or contractor. Architectural embellishments cost money in time and materials as well as design expense, particularly if an architect is not familiar with log construction. All in all, most log home owners should be able to get the house of their dreams without the services of an outside architect.

Acquiring Land

LOCATING LAND

For many potential log home buyers, the biggest obstacle has nothing to do with the house itself. Rather, it is finding the place to put their dream home. The problem seems to be greatest near urban areas. Land surrounding urban areas is generally more expensive and often subject to numerous zoning restrictions. Finding a place to build your dream log home may be a greater test of your patience than building the house itself. Approach your land search armed with a system and some solid background information, and you will find the experience less frustrating and even enjoyable.

Start by reviewing the desirable features of your home site. How much land do you want? Do you prefer wooded land or open, sloping or flat? Are you seeking a walk-out basement or basement garage? In what area should this land be located? Is distance from a workplace a factor? Use the Land Locator Checklist in the Appendix to define your property needs as specifically as possible. Remember though, flexibility in land requirements will make it easier to find a parcel and proceed with your plans.

Some land characteristics can affect your budget significantly, so bear this in mind as you start your search. If your land is in a development that has hard-surfaced roads, public water and sewer, and utility service or easements, some of your potential site problems have been eliminated. However, you may still have to pay tap fees or hook-up charges to gain access to those services. A phone call to the appropriate agency will tell you what costs will apply.

If you are seeking rural land, you will probably need to consider the cost of an entrance road, well, and septic system in your overall land budget. For estimates of well and septic costs, consult neighbors in the immediate area of the land you are investigating. They may be able to provide price estimates or refer you to well drillers and septic contractors. If neighbors are not available, consult the phone directory. County health departments sometimes publish a list of approved well and septic contractors. (In some areas, only licensed or approved contractors are allowed to install systems.)

Land clearing is another consideration. Even if you favor wooded land, you may need to have some trees removed. Costs for such removal will vary with the size and number of trees and how you want to dispose of them. Simply dropping the trees is considerably less expensive than cutting them into firewood lengths, stacking them, and grinding or digging out the stumps. Here is an excellent opportunity for you to put some "sweat equity" into your project. Just be realistic in estimating the time available and your physical abilities.

Probably the simplest method of finding land, short of inheriting it, is to consult a realtor. Outline your wants and budget and let a professional do the initial leg work. This works well if your time is limited or if you currently live a considerable distance from your proposed building site. Look for realtors who specialize in rural land. Local banks can often point you toward the local raw land specialist. Or simply review land listings in local newspapers and see whose name appears most often.

Whether you choose a realtor or not, don't discount the "grapevine" as a source of information. Not everyone lists land through realtors. Try to get acquainted with local people early in your land search — after all, they may become your neighbors. Inquiry at local restaurants, service stations, or farm supply stores can sometimes produce a lead for land available for substantially less than that listed through realtors. But beware of unusually good bargains. Buying raw land contains an element of risk, so research each potential parcel thoroughly before taking any action.

The main risk in buying rural land comes because not all rural land is "buildable." Perfectly healthy-looking countryside may be completely unsuitable for your purposes for a variety of reasons. First, there are legal reasons. Does clear title exist for the land? That is, can current owners produce legally acceptable proof of ownership? Are there zoning requirements or covenants that would prevent you from building the kind of house you want or restrict the way you site the house? Then there is the land itself. Are there springs or drainage problems that would prevent putting in a foundation? Is there concealed rock that would make it necessary to use explosives to put in the foundation (expensive!)? Finally and perhaps most important, does the land "perk"?

A "perk" test is a simple test done to determine whether the soil is suitable for a septic system. In many areas of the country, no perk — no house, period. A perk test is done by digging a hole (or holes) in the proposed septic area. Depth of the hole varies according to local regulations but usually ranges from a couple of feet to no more than 12 or 14. The inspector will fill the hole with water to a predetermined depth and then record the amount of time required for the water level to fall a specified amount. The water level must not fall too quickly or too slowly.

In some areas, there are alternative sewage systems allowed on land declared unsuitable for conventional septic systems. In my area, "sand mound" septic systems are sometimes approved if the land fails a conventional perk test. A sand mound system is similar to a regular system except that special sand is brought in to construct the drainage field. Sand mound systems are not allowed in all areas; their cost is two to four times that of a standard septic field, and they can be noticeably unattractive. It's best to understand your sewage system options before you make a commitment to buy a parcel of land.

Increasingly, especially in developed areas, building permits will not be issued without a satisfactory perk test on record. Most land shoppers restrict themselves to land with an approved perk site. In cases where a perk test has not been done, I recommend a contingency in the land contract that makes the final sale subject to a favorable official perk test. A preperk test, an unofficial test conducted by a septic contractor, can be a worthwhile investment to determine whether a piece of land is worth pursuing. If land is unperked and the seller is reluctant to allow a preperk or a perk contingency clause — beware. You may be on your way to purchasing a beautiful campsite that will never hold a log home.

Be sure to check the status of the boundary survey of any land you consider. Make sure the seller actually owns the land being offered for sale. This is a particular problem in the eastern United States, where land records go back several hundred years and are based on "metes and bounds" descriptions. A typical boundary description may read: "Commencing at the large rock beside Yoohoo Spring and going southwest eleven perches to a white oak tree, thence northeast fifteen perches to a steel rod imbedded in the trunk of a large maple tree. . . ." Such a description was sufficient in 1750, but try to find the maple tree with the iron spike in it today! See if there is a recorded plat in the county recorder's office for the land you are considering.

When my company purchased land for our display home, we purchased through a real estate agent representing an absentee owner. There was a recorded deed, but no survey. The deed description seemed clear enough, so I went out with a compass and laid out the approximate boundaries of the property. The land lay on a wooded hillside with an ancient rock slide running down one side of the property. Just off the opposite boundary, a long ridge angling toward the property ended in a jumble of boulders just short of the boundary I had laid out.

"Sure glad that's not included—I'd hate to have to work around that," I told the realtor. We proceeded with the purchase. The county required a survey before issuing a building permit, so we hired surveyors and gave them a copy of the deed description. Two months later they called to report that they could not certify two of the boundaries.

"They don't match anything in the county records," the surveyor said. "You'll probably have to go through the courts to get clear title—happens a lot in these mountains."

"But what about the deed description?" I said. "I found one of the corners, and the compass bearings are clear."

"Yes, they are," the surveyor replied. "However, they're wrong. There's nothing in the tax records fitting your deed description, and recorded surveys of adjacent properties don't have matching boundaries."

Still unsatisfied, I located the surveyor who had originally surveyed the farm from which our parcel originated. After six months, he called to report a completed survey. I raced to his office to pick up the map. All the boundaries were shifted, and there in the middle of our land, rearing its stony head, was that jumble of boulders.

"But this doesn't match the deed description," I protested.

"I know—the deed description was wrong. Happens all the time in these mountains."

We had a contingency in our contract that required the seller to absorb the expense of the survey, but we still owned a rock pile, atop which now sits our display home, Laurel Lodge.

The moral of the story is, "When buying raw land, verify everything." And use contract contingencies to give you a way out should things not turn out the way you plan.

ROADS, WELLS, AND SEPTIC SYSTEMS

Most of us take a lot of aspects of urban and suburban living for granted. The city dweller moving to the country often requires re-education as well as an attitude adjustment. The three most basic services—transportation, water, and sewer—take on a whole new meaning when it's up to you to obtain and pay directly for what you need.

Let's start with access. Most likely, you will require a driveway of some sort. A driveway on a small building lot is not usually a difficult or expensive undertaking. But if you are building on the far side of a 20-acre tract, "driveway" may be an understatement. In such a circumstance, you will need an honest-to-goodness road.

A good road should be passable in all weather, require relatively little maintenance, and be able to handle everything from passenger cars to delivery trucks. You may also wish to consider the trucks that will be bringing your log home package. Generally, these are tractor-trailer rigs that may exceed 60 feet in length. Their turning radius and grade requirements are very different from a short delivery truck.

In laying out your road, consider the traffic that will use it and design it to accommodate the least flexible. If you can't or don't wish to accommodate large tractor trailers, prepare to reload your log home kit onto smaller trucks or use a fork lift. Remember to consider width, too. Trying to snake 20- to 40-foot logs between trees that are 16 feet apart is a challenge most forklift operators (and your pocketbook) can gladly forego.

Walk your property with a road contractor and get rough estimates of the cost even before closing on the land purchase. Good roads can be an expensive proposition, so it's best to know before you commit yourself how much you are going to have to add for road costs. Understand, too, what type of construction is used for the price you are being quoted. The road at Laurel Lodge consists of a roadbed cleared and compacted by bulldozer, covered with several inches of 2-inch stone, and topped with 4 inches of compacted, crushed rock ("crusher run"). Such a road will cost considerably more than just a couple of inches of crushed rock spread on the ground surface—but it will stand up to much more abuse and need re-grading and repair a lot less often.

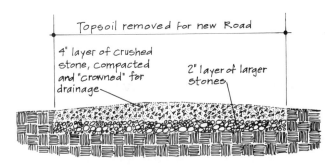

Topsoil removed for new Road

4" layer of crushed
stone, compacted
and "crowned" for
drainage

2" layer of larger
stones

ROAD CROSS SECTION

Wells are another aspect of country living. A well system actually consists of the well itself, usually with a well casing for at least a portion of its depth, a well pump, and a pressure tank. The pump resides near the bottom of the well. Flexible water line runs from the well to a pressure tank located inside the house. The pressure tank maintains a reservoir of water under a preset pressure. Water drawn from the tank by the plumbing fixtures in the house triggers the well pump, replacing the water that is removed. An undersized pump or pressure tank can cause temporary water shortages whenever water demand in the house exceeds the ability of the pump-tank to replenish itself. Low water yield can also result in temporary shortages. Watering a lawn, for example, on a hot day in late summer, when the water table is already low can cause the pump to draw all available water out of the well. Then you must wait several hours or several days for the well to recharge itself.

It's a good idea to investigate the water situation in the area you plan to build before you commit yourself to a piece of property. Ask neighbors about their wells and any specific water problems. They can give you a good idea of the depth of well, flow rate, and quality of water you can expect. If well water in your area is acidic, has a high level of dissolved minerals (hardness), or has a high iron or sulfur content, you may want to budget for a water softener, filter, or iron exchanger as part of the cost of your well system. In my area, many people have their well drilled as soon as their road is complete; a temporary, generator-operated pump can then be used to provide water during construction.

One word of caution about wells. Investigation of nearby wells is a good general indicator, but is not always accurate. Patterns of water flow underground may result in your well having characteristics completely different from wells located on neighboring property. Water from the well at Laurel Lodge is acidic and high in sulfur and iron. Two houses located within a quarter mile on the same mountainside have none of those characteristics.

Septic systems also go with country living. Basically, a septic system consists of a tank and either a drainage pit or drainage trenches located in a drainage field. Waste water flows from the house into the tank. There,

PHOTO BY JIM COOPER

Wells are a fact of life not always familiar to people from urban or suburban backgrounds. Costs will depend on the depth of drilling, and water quality can vary. Consult neighbors for a general idea of water quality and well depth in an area, but be aware that subterranean geology can vary considerably even over short distances. Well rigs will need a basic access road. You can minimize environmental impact by using the access route that will become your driveway.

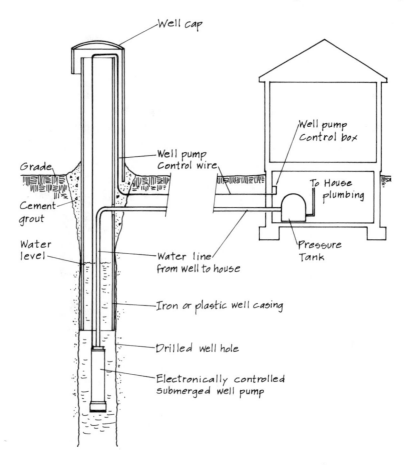

ANATOMY OF A WELL

solids settle out and are broken down by bacterial action. The waste water flows into perforated pipes through which it is returned to the soil. Operation of a septic system has several important implications for home builders.

First, waste systems are designed to use gravity flow whenever possible (in some areas this is a requirement). Where allowed, sewage injector pumps can be used to pump sewage against a gradient. This is naturally more costly and raises the potential problem of pump failure. Whenever practical, arrange your building site with the septic system down slope from the house.

The septic system should be located away and down grade from the well. In this area, county requirements dictate that the well be at least 80 feet from the closest point of the septic system and on a higher contour. Distance between the house and septic field is less critical. A well-built system should have no odor and be generally unnoticeable. In many areas, there are restrictions concerning crossing a septic field with a roadway. In any event, it's a good idea to avoid anything that could compact the soil in the septic area.

Some areas have alternatives to traditional septic systems. They may allow holding tanks in places where soil is unsuitable for a drainage field. In these cases, a large tank is buried near the house to receive waste. Periodically, a waste hauler must pump out the tank. Sewage lagoons are

PHOTO BY JIM COOPER

Septic systems are another fact of country life. This concrete septic tank is being buried at the corner of the septic field. It will receive waste from the house which, after a period of settling and decomposition, will flow into trenches or pits in the septic field to be absorbed by the soil. Like well-drillers, septic installers need cleared access.

allowed in some areas, although these are generally designed to receive waste from several houses or a small housing development. Composting or "waterless" toilets may be desirable in certain situations. The final determination will be local regulations and budget.

SITE PLANNING

An important consideration in terms of both budget and enjoyment of your log home is the way the house will be sited on the land. While remote or inaccessible sites may provide beautiful views, seclusion, and serenity, they often come with a price tag. Long driveways, foundations blasted out of solid rock, and log kits that must be off-loaded from tractor trailers then reloaded onto smaller trucks to negotiate narrow or winding access roads all can add substantially to the cost of a log home project. Also, sites where materials must be stored inconveniently or at a distance from the construction site will take longer to build, resulting in higher financing costs and possibly greater labor costs as well. When you are evaluating potential locations for your log home, consider not just the finished product but the construction process as well.

Difficult sites are not always remote. I once built a log house that was to include a precast foundation. The house sat atop a small knoll beside an

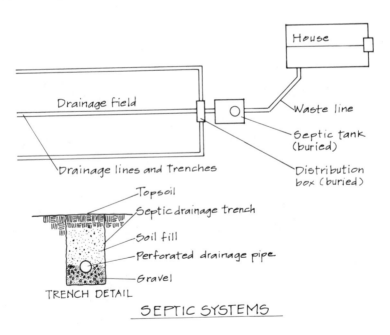

House

Drainage field

Waste line

Septic tank (buried)

Drainage lines and Trenches

Distribution box (buried)

Topsoil

Septic drainage trench

Soil fill

Perforated drainage pipe

Gravel

TRENCH DETAIL

SEPTIC SYSTEMS

PHOTO BY JIM COOPER

Efficient construction requires efficient material handling. Designate a storage area near the house site for log storage. Many companies expect the homeowner to be responsible for unloading log trucks. This will require a forklift and an operator. The extendable-boom rough-terrain forklift shown allows faster unloading under a wider variety of field conditions than the conventional mast-type forklift. Rental costs for the two types are similar.

asphalt road. The drive was one small hump not more than 50 feet long with an elevation change of about 15 feet. Trucks, bulldozers, and back-hoes had been running up and down the graveled slope for over a week. But when the crane came to set the precast foundation panels, it took two bulldozers to get it into position. I have pictures of that thirty-five-ton crane with one bulldozer pulling and one pushing, like circus workers trying to get a reluctant elephant up a ramp. With the wait for the un-anticipated bulldozers, I calculated the additional cost was over $500 and a half-day of work time.

Begin a site plan with a clear idea of the entire site. If possible, use a survey drawing or plat map that contains topographic information. Pay attention to the direction the house will face. This is especially important if you anticipate a lot of window glass in one area of the house. Try to face the house to gain maximum advantage from the path of the sun. Generally, this means facing the house south or quartering toward east or west. Placement will have an impact on energy efficiency. In unusually hot or cold areas, you may want to site the house to take advantage of shade or windbreaks.

In many areas, local regulations will have something to say about where you can place your house. Setbacks are distance requirements from boundaries and public features such as roads. Easements are areas through which access must be allowed. They may be public, such as utility easements to allow utility lines to cross your property or allow access for utility company maintenance equipment. They may be private, such as a road easement that allows an obstructed neighbor access to his or her land. Easements are usually identified on deeds and pointed out in sales contracts. Information on setbacks may have to come from planning and zoning offices or from building permit authorities.

Major restrictions on building location are based on the location of the well and septic system. In many areas, your building site must be approved by the local health department before a building permit can be issued. Building plans that include septic fields on steeply sloped ground, in floodplains, near permanent waterways, or in areas of major surface or sunken rock, may all be deemed unacceptable. Find out early in your site planning which restrictions can affect you. A call to the local health authorities or building permits department, or a visit with local septic contractors and well drillers can provide you with the necessary information.

One approach to siting your house that's useful for both estimating and visualizing the finished product is to stake the house, well, and septic locations. Show these areas to officials and prospective contractors. This can help you anticipate and avoid potential problems.

If you do not have topographic information from a surveyor on your house location, you may want to examine the site with a builder's level or transit. These instruments can accurately show how much of the basement will be exposed, whether retaining walls or frost footings will be required, and a number of other considerations. Eyeball estimates are notoriously inaccurate. One customer of mine, planning to build on a sloping site, was skeptical when I said that the front of his house would have 4 feet of foundation exposed and that the attached garage should probably be dropped several steps to avoid having to build a retaining wall and put the driveway on fill dirt. He simply refused to believe me until I brought a builder's level and showed him. (Actually, he even refused to believe me then. He watched me set up and operate the level. Then, after I left, he and his wife rented a level and repeated my measurements. "But, it looks so much flatter," was all he could say when I arrived the next day to lay out his house.)

FINANCING LAND

For conventional home owners, buying land and buying or building a house usually are covered by a single financial transaction. Because log home owners generally seek rural settings and land parcels rather than building lots, financial details get a little more complicated. Finding and purchasing land often is a process all its own.

Generally, there are three ways to obtain land: (1) inherit it, (2) steal it, and (3) buy it. If you are fortunate enough to fall into category one, congratulations, you have jumped one of the taller hurdles on the path to your dream home. If you fall into category two, probably you now enjoy (or soon will) complimentary housing (unfortunately, not log housing). For most of us, however, category three is our only option.

There are two basic ways to buy land: (1) pay cash in full, or (2) use financing. The latter category covers most log home purchasers. Land financing has some similarities to and some differences from home financing.

Your land purchase may be financed by a lending institution, the current land owner, or a third party (individual or corporate). Although most land purchases are financed by lending institutions, you may have to do some research to find which ones are willing to lend on land only. Many lenders are interested in lending only on a land-house package. Because land values are so subjective — one person's home site is another's howling wilderness — lenders face greater risk in loaning money on land alone. The more remote or unusual your tract of land, the more difficult it will be to get lender support.

As part of your shopping for log home financing, you may want to ask prospective lenders if they will finance a land purchase separately. Get a list of requirements to make sure you can qualify. Before approving a land loan, a lender generally will require proof of clear title and an appraisal. In addition, you will be expected to make a down payment of at least 20 percent of the purchase price. The lender also may require proof that the land will support a septic system and well.

It is in your interest to obtain a title search to verify ownership of the land. The search can be obtained from an attorney (expensive) or a company specializing in searches (cheaper). Your lending institution can recommend someone and may even make the arrangements for you. The only thing you can count on is—you'll get to pay the bill. Some lenders require title insurance as additional protection against possible ownership problems ("clouds on the title").

A land value appraisal is also in your interest, because it places a realistic market value on the land you are investigating. A market value much lower than the asking price of the land may give you a basis to ask for a price reduction from the land's seller. Because land purchases in many areas of the country hinge on institutional financing and because lender guidelines generally restrict them to a fraction of the market value, appraisals can exert considerable leverage in price negotiations.

As with a home or construction loan, you will be expected to demonstrate an ability to repay the amount you are borrowing. The lender will require employment records, proof of income, and a record of outstanding debt. The term for a land loan is generally less than for a home loan. Payments are usually based on periods ranging from five to fifteen years.

Financing is discussed in more detail in Chapter 6.

Defining Your Goals

THE BEST PLACE to begin your log home project is with a clear understanding of your personal objectives. The clearest path to disaster in any building project is to enter into it with a collection of fuzzy notions of what you want. A couple who were early customers of mine came to me with a vague idea of the log home they wanted to build. They brought me a picture from a magazine with instructions: "Something like this — sort of," and a vague budget. My company prepared preliminary floor plans from their rough sketches. These were modified, then rejected and another "general" design took their place. More plans, another estimate, more modifications, and eventually another rejection. . . . After a year, seven house plans, and estimated project budgets ranging from $225,000 to $750,000, they finally told me that they were not going to use my company because we moved too slowly and lacked the expertise to give them what they wanted! In truth, they needed someone to *tell* them what they wanted.

If those customers had spent some time outlining their goals and making firm budget decisions before coming to me, they would have been sitting in their log home a year later instead of just starting the process with another company. To take your log home from dream to reality and to have the reality most closely match the dream, it's best to start with a clear set of goals. Goal-setting can be divided into several stages: (1) a rough idea of the scope of your project, (2) a preliminary cost analysis, (3) a hard look at your financial and physical limitations, and (4) a priority ranking of your desires.

Let's start with the scope of the project. This step is fun because here is where you get to dream. Later we'll apply the hammer of reality and see which parts of the dream can withstand the blows. Right now, I recommend investing in a sturdy three-ring notebook, some lined paper, and a set of notebook dividers (don't forget to save the receipt). Label the first notebook section "Goals." Now make yourself comfortable and start dreaming a log home.

In dreaming your log home, try to visualize as much as possible. Where is it located? What kind of setting and how big a lot? What style of house? Does it have large squared logs with dovetailed corners and stripes of white chinking, or are the logs round and tightly fit with crossed corners? Is the house one story or two? Are there porches or decks? When you step inside the house, do you see soaring cathedral ceilings with exposed beams? Are the walls inside covered with wood or drywall?

As you dream, note the features of your log home in your notebook. List as many characteristics as you can. The checklist of "Dream Features" in the Appendix can help. The dreaming process can take an evening, a month, or a year. If this is to be a family project, get everyone involved. Use this list to help define your dream log home, adding any special features you might want that aren't listed here. Collect pictures or sketches to show how you would like various features to look. Use the list and illustrations to show log home sales people and subcontractors how you want your home to look. This will help insure that your house turns out the way you imagine it, and it helps you budget by giving log home manufacturers and subs enough information to provide accurate cost estimates. When you have pared your list of dream features to match your budget and site considerations, use the list as a definition of the scope of your project.

THE FLOORPLAN

There are many sources of good log home floorplans. The plans appearing in log home company sales literature are a good place to start. These are usually tested plans that may be comparatively less expensive because blueprints and materials lists already exist. But you needn't restrict your research to log home catalogs. Books, magazines, and even floorplans of friends' homes can contribute to your own home design. Make notes and take pictures of features you like.

A good way to review the practicality of a floorplan is to do a mental "walk through" of the house. Study the plan as you mentally arrive home from work or shopping. Enter the front door: What do you see? Is there a coat closet handy? Are you standing on carpet or hardwood, or is there a separate floor area to catch the rain or snow that falls from your coat? What is the impact of muddy feet? Imagine entering your log home burdened with groceries. Do you have to climb several flights of stairs or walk across the living room to get to the kitchen area? How close is the nearest cabinet where you can unburden yourself? What happens when you step out of the shower and find there's no towel on the rack? Is there a linen closet handy, or do you hop naked down a long hall hoping that your daughter's friends in the living room don't suddenly turn around? Imagine your life as you want it to be in your log home and see how the floorplan fits.

Use your prospective floorplans with your site plan to determine the view from various places and the placement of porches and decks. You may find that one floorplan orients the living room windows toward a spectacular view, while another plan ignores it. Such mental exercises will help you get the most out of your design and can help anticipate difficulties and uncover potential costs.

Here are some considerations in developing your floorplan:

Entries

Entry areas, especially when they open from an unprotected exterior, benefit from resilient or grime-resistant flooring. Consider sheet vinyl, tile, or slate to provide an area to remove wet overcoats and boots before stepping onto hardwood or carpeting.

If you like outdoor activities such as hunting, fishing, or gardening that can involve soiled equipment and clothing, you may want to consider a mud room off of one entry. The mudroom may contain a utility sink and storage shelves, and may double as a laundry room. Whether the mudroom is near the main entry, rear entry, garage entry, or basement entry depends on your lifestyle.

Coat closets or furniture that doubles as clothes storage near entries help control clutter.

Kitchen

Does the kitchen have easy access from the outside? A double armload of groceries gets heavier and more unmanageable the farther you have to carry it and the more corners and stairs you have to negotiate. What about kitchen layout? Many people come to me with elaborate, large kitchens. They look great until I point out that the refrigerator is a half-day hike from the sink. There are many good books on kitchen design. Using one can help you avoid an awkward, impractical layout. Consider a pantry in or near the kitchen to avoid carrying canned goods from the basement.

Baths

Baths are a current fancy of home designers and builders. Make sure there is at least a half-bath easily accessible for guests. I have used a design that allows a bath to be used as either a full or half-bath. The vanity/toilet area may be entered from a hallway, while the shower/tub area is entered through a bedroom. A door separating the bath area from the vanity/ toilet area allows use as a public bath, while the bedroom entry allows use as a private full bath when desired. This arrangement works well with guest bedrooms when there is a need for a second master bedroom (such as when couples share a house or when several generations of family members are living together). Baths are also a potential money pit. Consider your lifestyle when laying out elaborate bathrooms. I've seen a number of baths with enormous tubs, whirlpools, and spas that haven't been used since the house was occupied. They look inviting in photographs, but are you someone who will spend forty-five minutes a day soaking? If you plan to have a full basement or unfinished attic, you may want to "rough in" a bath in this area. A rough-in simply gets the necessary basic plumbing in place, with no fixtures, finished floor, or walls.

Your Land

The land you intend to build on can affect some of your design decisions. For example: You've inherited a parcel of land and your dream is to put a log home on it. One of the "must-have's" in your log home design is a walk-out basement. Unfortunately, the land you inherited is flat as the proverbial pancake.

There are two ways out of your situation: (1) Sell the land you've inherited and use the proceeds to buy land compatible with the house you desire, or (2) modify your log home desires to accommodate the land that you already have. You might, for example, elect to replace the walk-out with a second-floor loft with an exterior balcony. If you already own land, it should be considered in the type of house that you design. If you don't own land, you can design your house any way you wish, but recognize that your design may limit the selection of land that is available to you.

My suggestion is this: If you already own land, show it (or a topographical map of it) to your log home representative at the beginning of your design process. If you don't own land, tell your log home representative to alert you to any design features you select that could affect your land purchasing, as you design your house. This can save you both time and money.

SETTING PRIORITIES

With a complete, concise list of the details of your dream log home in front of you, it's time to set some priorities. Go through your list and rank each feature as "MH" (must have), "LAL" (like a lot), "SWBN" (sure would be nice), and "WITAMLO" (well, if there's any money left over). Notice that with my abbreviation system, as priority decreases, the abbreviation becomes harder to remember and takes more time to write. Don't be lazy and just make everything "MH."

Now it's time to review your limitations. There's little point in investing more mental exercise and emotional energy in producing a dream project that is beyond your budget or building circumstances. In addition to your notebook, you'll need a calculator and a telephone for this step. Start a new page in your notebook labelled "Dream Home Budget."

Starting with the lot, try to put a price tag on the features that you have noted on your dream list. If you own land, list its value. If you plan to buy a lot, check with a realtor or newspaper listing to get a rough figure for the size and kind of lot you are looking for in the area you want to live. If you are flexible about location, size, or type of building site, list a cost range based on the types of variation you will accept. If your building site will require utilities—water, septic system, electric service, telephone, cable TV—call for rough estimates of how much each of these will cost.

Well drillers and septic system contractors usually are listed in the telephone Yellow Pages. A call to the county planning and zoning office can give you information on the availability of electric, sewer, and water service in your area. The utility company can give you rough estimates of the cost to provide service to rural lots.

When you get to the cost of your house, a call to a custom builder or a log home builder in the area where you will be building can give you a rough figure for the cost of your log house. Provide a basic description of the house—size in square feet, number of baths and bedrooms, number of stories, decks, porches, garage or carport—and ask for the cost of a similar house (not necessarily log) that the company has built recently. Get several figures, if possible, and show the range on your Dream Home Budget page.

As much as possible, avoid the use of "cost per square foot" estimates from builders and log home dealers. Every house, log or otherwise, has a unique square-foot cost associated with it. This cost is obtained by building the house, taking the final complete cost, and dividing by the square footage. Many factors contribute to the square-foot cost, ranging from building permit costs, through site requirements, to amenities such as fireplaces and whirlpool tubs. In the area where I sell homes, there is as much as a $2 per-square-foot cost difference from one county to another based solely on cost of permits and inspection fees. Simply by crossing the county line, substituting hardwood floors for carpeting, adding a full masonry fireplace instead of a woodstove flue, and putting a whirlpool tub in the master bath, the per-square-foot cost of a moderate-sized house can be varied by as much as $15–20 per square foot. (None of these variations affects the cost of the log home kit.)

View kit factor estimates and estimates by log home dealers in general with some caution. To say that a log home will cost two to three times the cost of the kit leaves a lot left unsaid. Does that two to three times include a basement? What kind of heating system? What kind of roof system? What location? A 20 x 30 log box on a slab, with a truss roof and fiberglas shingles, heated with electric baseboard heat, will show a turnkey cost (per square foot or kit multiplier factor) that will make your heart soar. But if you use that figure as a basis for estimating a 2,000-square-foot house with cathedral ceilings, heavy rafters, forced-air heat, and central air conditioning, you'll find that your estimate may be off by $30,000.

Be skeptical of any house price that places your cost lower than that of a similar conventional home. I'm not saying it isn't possible; I'm just advising caution. Also be careful not to base your comparison on the cost of a tract-built conventional home. A builder working in quantities of twenty to two hundred homes can generally sell those homes at a profit for less than the final cost on a single custom home.

When you have placed a price tag on your lot, site improvements, and house, add the figures to find your total project cost. Now it's time for a reality check.

You may already have an idea of what you can afford or want to spend. Compare that figure with what you've established as a house cost. If the project cost is within your budget, pat yourself on the back for being a rare individual and head straight for the chapter on getting under way.

If you need to determine what you can afford, here's a simple way to proceed. Make an appointment with the loan officer at your bank or at a financial institution that deals in home mortgages. You may want to inquire whether they have experience with log homes, although at this point it isn't really necessary. (I'm assuming that, like most people, you will need additional financing.) When you meet with the lender, take the following information: your annual income, a summary of outstanding debts, a list of your assets, and an estimate of the cash or equity that you have to contribute toward the cost of the project. (This can include the amount of cash you have invested in your lot, equity in your present home, any lot improvements you have made that are paid for, deposit and any additional payments made on your log home kit.) The lender will use this information to determine the amount of financing for which you qualify.

How do they match? If your projected budget is within reasonable range of your lender's prequalifying figure, there's a log home on your horizon. If the lender's prequalifying number is substantially less, better sit down with your priority list and start looking for places to pare your budget. You may want a second opinion on the amount your lender feels you can handle, but don't expect a lot of variation on this point from one lender to another. The formulas used for prequalifying are fairly standard throughout the industry.

To trim your budget, first look at your overall project cost as it compares to your lender's figures. A hard look should tell you whether there's hope at all for your log home in its present form. Be realistic! A veritable army of log home hopefuls has trooped through my door with a log house project that can't be built for less than double what their financial circumstances will allow. And many of them believe that the problem can be solved by simply eliminating the whirlpool and leaving off the deck. I wish it were that easy!

Instead of trying to perform financial magic tricks, sit down with your priority list and try to attach some realistic cost figures to the items on your list. Your log home dealer can help with items related to your log kit. But unless he will also be your builder or has direct building experience, it's best to talk with a contractor about items not directly related to the kit. My rule to follow is: To get the only realistic estimate of the cost of anything, call some people who sell it or do it and ask for a written quote. If they will attach their name to it, it's probably realistic.

Start totaling costs with the lowest priority items on your list. See how closely they bring you to your lender's figure. Maybe you'll bring your budget into line with your dream fairly intact. Throughout the process, keep an eye out for less expensive substitutions and items that could be added later. Decks, fireplaces, whirlpools, hardwood floors — all can add up to substantial savings if you're willing to wait.

If your reality check indicates that a simple budget trimming is inadequate, you have only two options: (1) wait until your financial situation changes; or (2) modify your dream. Ask yourself which is more important: living in a log home or living in a *specific* log home. With the knowledge you have gained from your research, sit down with your log home dealer or several log home dealers and discuss what will fit your financial situation.

CONTROLLING COSTS

I would like to be able to present cost figures or ranges to help you budget, but geography, local labor conditions, land prices, and regulatory conditions make cost generalizations impossible. As I write, I could buy a finished new house including land in Salt Lake City for approximately one-fourth of the cost of a similar house in my area. (The cost of the materials for the house alone is actually higher here than the cost of the completed home on land there.)

One thing that seems to be difficult for many log home shoppers to grasp is that a turnkey log home is almost never cheaper than a conventional home of the same size. Although most manufacturers and sales reps will state this up front, a few do it with a wink, as if to suggest, "I have to say that, but we both know it isn't so." A few still say that log homes cost less than conventional homes. If you encounter this view, proceed with caution. The claim is often made like this: Mr. and Ms. Log Home Shopper select a 2,000-square-foot, three-bedroom, two-bath house. It has two floors with a loft and a full unfinished basement. Our shoppers have visited several conventional home developments and have seen similar homes that have price tags of around $100,000 when land value is deducted. Based on this, they believe their log home should have a turnkey cost of around $90,000–$110,000. They think they are basing their estimate on an accurate comparison. After all, the conventional homes are the same size and have the same floorplan and are in the same general area where they want to build. But is the comparison really accurate?

Consider their log home. It has open beams supporting the second floor and heavy timbers supporting solid wood ceilings. Most of the walls are tongue-and-groove wood rather than drywall, and the interior trim is all stained wood rather than painted, the doors are solid wood rather than hollow-core, and the stairs are custom-made on site from solid wood. The kitchen cabinets are solid wood with raised panel doors, and many of the floors are hardwood. The eager shoppers concede that this could make a little difference in price but would hardly expect the $20,000–$30,000 that it would cost to put those same features into a conventional home.

In addition, consider the economy of scale—the fact that the developer who built the conventional homes they were using for comparison started with a 200-acre tract of land and subdivided it, thus allowing a generous profit at less than the basic cost of a single tract of land the same size. (Most log home shoppers are not interested in quarter-acre tracts.) The developer installed water and sewer lines, dividing the cost among four hundred home sites. The log home buyer will have to bear the cost alone of a separate well and septic system. Finally, the log home buyer is paying for materials and labor on a one-time-only basis and using subcontractors to boot. The developer is buying lumber by the train-car load and paying wages to workers far below what a subcontractor must charge to be profitable. In the final analysis, whether it comes from a kit or a handcrafter, a log home is a custom home and will cost, as with any custom home, according to the features that it includes. If someone offers you pricing for a log home that seems well below the price of conventional housing, the first thing you should ask yourself is, "At these rates, why do I not see log homes everywhere, and why are developers not abandoning conventional housing?"

I don't mean to discourage log home enthusiasts. A well-built log home is a thing of beauty that, to my mind, no conventional home can equal. In addition, kit home or not, it is built log by log and has the potential for a level of quality that simply cannot be found in most conventional housing. I simply believe that if you are preparing to sign a thirty-year mortgage and commit the major portion of your income to a house, you should understand, before the papers are signed, exactly what to expect. I have heard too many tales of people "running out of money" and winding up selling their dream house before it's even complete. I also know of more than a few log home builders who have gone belly up because the final payment wasn't there when the job was finished.

With that said, let's look at some ways to control the cost. There are basically three factors that can affect the price of your log home. (I'm talking about the house only—not land or sitework.) These are: (1) size, (2) design, and (3) labor.

Size

Consider carefully the size of the log house that you want to build. Base room sizes on rooms that you have seen or lived with. I talk with many people who have sized their house according to the latest housing trends rather than using any practical basis. When you are determining the size of your house, remember that you are going to have to pay for the space four times: you have to build it, you have to furnish it, you have to heat and cool it, and you have to clean it. For most of us, there are a lot more interesting things to do in life than housecleaning or struggling to earn the money to pay for a large house, lots of furnishings, and hefty utility bills.

Design

The design of your house can greatly affect your house price. Large open areas with cathedral ceilings are beautiful but expensive to heat and cool. Because work proceeds more slowly and often must be done from scaffolding, cathedral ceilings also carry higher labor costs. Complex shapes and roof lines may look interesting, but they come with higher material and labor price tags than more basic profiles. Usually, complicated designs also mean reduced living area for the size of the house. A customer came to see me with a floorplan from another company. He wanted my log package and a price from me for a turnkey house. When I gave him the figure, he was shocked, "But the house only has 2,600 square feet." I pointed out that, while that might be the living area, the actual area that I had to consider for construction purposes was closer to 3,200 square feet. His design had a complex roof line, and a considerable amount of second-floor area was lost under sloping ceilings that nevertheless required finishing. The simpler the profile of the house, the less expensive it should be to build.

Two-story houses with a given square footage of living area are less expensive to build than single-story houses of the same square footage. "It's cheaper to build up than out" is an expression used by many builders. Building a house "out" means additional excavation, basement, and roof. Building "up" means more living space within the same basement and roof with only additional walls and floors. Heavier heating and cooling requirements also go along with a house that spreads out rather than up.

Most subcontractors vary their pricing according to the complexity of the design, figuring that complexity means more time and contains greater possibility of estimating errors. When considering carpenter subcontractors, be skeptical of anyone who bases a bid on the price of the log home package. Imagine two log houses with the same kit price (and even the same approximate square footage). One is a simple rectangle that contains many windows, skylights, and several dormers. The other house has a complex shape with no skylights and a modest window package. Both log kits cost approximately the same. According to a subcontractor who charges according to kit cost or square footage, both houses would cost the same. In reality, the price of the simple house has been raised because of the window package. It should actually take much less time to dry-in and thus should carry a lower labor cost. The cost of the house with the complex shape is driven by the log cost; it should take longer and cost more. In my experience, dry-in prices based on kit prices or square-foot estimators tend to penalize people with efficient designs and benefit people with more complex designs. (As one builder told me, "We charge according to the kit cost. On some houses we make more money; on others we make less.")

Labor

Keep in mind that, whether the labor prices you get are for an entire project, by the square foot, or by the hour, the subcontractor's main concern is going to be time requirements (plus any special expenses like equipment rental). I use a computer estimating system of my own design that considers the factors I have encountered that can affect the time required to erect a log kit. This way I can price every log house fairly and accurately. You may want to show such a form to potential carpentry subcontractors. It can protect both you and the subcontractor by helping you both understand the amount of work involved in your particular project. The Carpentry Labor Estimator form is reproduced in the Appendix.

ESTIMATING COSTS

Construction budget items will fall into three categories: materials, labor subcontracts, and subcontracts that include both labor and materials. Use the Cost Estimator Checklist in the Appendix to estimate the total cost of your construction project. I use a similar form to cover almost all house-building projects. Study the form carefully and add any categories necessary for your special requirements.

The figures you will place in the form will be either firm bids or estimates. Obviously, the more firm prices you can obtain, the more accurate your overall estimate. However, you may want to start with rough estimates simply to determine whether you can afford the project. That's fine, as long as you distinguish between estimates and firm figures. I complete the form using a pencil and placing parentheses around estimated figures. When I have a firm number, I simply erase my estimate and replace it with the quoted price. A quoted price, by my definition, is a figure that someone has signed his name to, saying that he will perform that work or sell those materials at that price.

To get realistic estimates, call suppliers or subcontractors in the category you are estimating. Describe your project and tell them you are just looking for a rough figure. They can usually quote you a range or ballpark figure. It's best to verify numbers by calling a second source. If the two figures differ by more than 10 percent, call a third and maybe even a fourth, using a middle or higher figure, or averaging, for your estimate. For greatest accuracy in your estimate, call suppliers and subs that you might eventually be using.

Now your rough estimate is complete. You have confidence that the figures on your sheet are realistic. The price is within your budget, even allowing a 5–10 percent margin of error. Your estimate can now be used to provide your lender with information necessary to process your application for financing.

Bank policies vary. Some provide their own budget sheet; others will accept the numbers submitted on the sheet you've already completed. But virtually every bank will require some proof of the accuracy of the numbers on your sheet. This proof comes in the form of a signed estimate or proposal from your log home manufacturer for the kit that you've selected and signed bids or proposals from the subcontractors and suppliers that you've chosen to use.

For items like appliances, lighting fixtures, and other finish details, you can generally enter an "allowance" figure instead of an actual cost. Simply sitting down with a catalog or visiting local stores can generally give you enough information to calculate an allowance. The bank construction loan department will judge its accuracy. I've noted on the estimating sheet those items that are generally submitted as allowances.

6

Financing

A MAJOR CONCERN for most log home buyers is the availability of construction and permanent financing. When people ask me whether they will have difficulty financing their log home (assuming they meet the financial conditions for the amount of money they want to borrow), I tell them that they may experience difficulties, depending on their circumstances. As log homes become more common, banks are finding that, despite their idiosyncrasies, log homes generally carry low risk to lenders. In most areas, at least one or two banks are probably experienced in log home financing and will be willing to work with prospective owners.

Three areas related to the uniqueness of log construction cause many financial institutions to be reluctant to lend on log homes: (1) determining appraised value, (2) paying for the log kit on delivery, and (3) handling owner-builders and owner-contractors. Here, from the bank's perspective, are the three problem areas. If you understand the bank's concern, you may simplify and speed the financing process for your own log home.

In order to secure financing, lending institutions must assign a value to your project. They will then lend a portion of that amount, depending on factors like economic conditions, your financial situation, and their perceived risk. Generally, lenders use established "loan-to-value" ratios (LTVs) to determine the amount of money they will lend. A new home's value is established by comparing it to existing similar homes. An appraiser studies the blueprints and the land, then compares your proposed log home with several similar homes to produce a "comparable value appraisal." Comparable value appraisals depend on recent sales data of similar homes. Do you see the problem that's coming?

There aren't that many log homes in any given area (lenders usually want "comps" from a small area—a ten- to twenty-five mile radius, for example). Log home owners tend to seek unique houses, and they tend to stay in them once they are built. What are the prospects of obtaining sales data on two or three log homes of a design similar to yours, sold within the last six months, within twenty-five miles of where you plan to build? Not very likely.

Lenders face the comparable value hurdle in different ways. Some simply refuse to lend. Others rely on appraisals that undervalue the log home. They reason that, since log homes are "unusual," they will have fewer potential buyers. They give the log home a lower appraised value than a similar-sized conventional home. This burdens the log home buyer to make up the difference with a larger down payment. But more data is becoming available to educate lenders. A nationwide study by the National Association of Home Builders' Log Homes Council has shown that log homes sell for prices similar to comparable conventional homes. As that awareness spreads, the comparable value hurdle will become less significant.

A second hesitation by many lenders concerns payment for the log kit. Many manufacturers specify that the balance of their kit must be paid for upon receipt—usually by certified check—before materials are unloaded. Lenders are not in the habit of releasing part of the construction loan to cover materials. They normally release money only upon completion of certain phases of the construction. Thus a lender may not release funds from the construction loan until the log package has been erected.

There are several ways to overcome this obstacle. Some manufacturers offer interim financing. Others will accept an "irrevocable letter of credit" from the lender. This is an official document that tells the manufacturer that funds from the construction loan have been placed in escrow specifically for the log kit. The letter must state that funds will be released directly to the log home manufacturer when the house reaches a certain stage of completion. However the bank deals with releasing loan funds to pay for the log kit, the time to find out is during the loan application process—not the night before the kit arrives.

Owner-builders and owner-contractors present a third potential pitfall for log home financing. More lenders now require the name of a licensed general contractor as assurance that a job will be completed and that the finished house will have a value that exceeds the loan amount. Part of this trend results from bad experiences by lenders whose customers could not build the house within their projected budget or whose workmanship was substandard. An example would be a homeowner with limited building experience who underestimates the cost of the house and also finds that he cannot complete the house within the time period of the loan. At the end of the loan period, his money is gone, he can't afford to continue the interest payments and, because of poor-quality workmanship, the appraised value has fallen, so the lender cannot risk increasing the loan amount. The outcome is that the lender will foreclose on the loan, taking the log home which probably must be partially rebuilt before it can be finished and sold (probably at a loss to the lender).

Unfortunately, this situation is not as rare as it should be. In trying to locate potential lending sources in my area, I have spoken with several lenders who will not even consider an owner-built or -contracted log home project because of exactly this experience. There is currently, within ten miles of me, a log house sitting vacant and unfinished, waiting for someone to pick up someone else's shattered dream.

If you are planning to act as your own general contractor, your prospects are better than those of an owner-builder. The owner-contractor has to sell the lender on his ability to manage a project, rather than his carpentry, painting, and masonry skills. With signed subcontracts from professionals making costs more certain, lenders feel less risk and are often more flexible.

If a lender will not consider an owner-contractor, there is still some hope for owner participation. The lender may accept a construction manager. This is a professional contracted by the owner to see that the work gets done properly and in a timely fashion. The home owner is still responsible for the bidding, subcontracting, scheduling, and payment of subcontractors but will be assisted and supported by an experienced professional. Local builders' associations, log home manufacturers and reps, architects, and phone directories are sources for a professional construction manager. Payment is usually a flat fee or a percentage of the final home cost.

If you want to act as your own contractor, but have limited experience with home construction and will be working a full-time job while your house is under construction, I strongly urge you to consider a construction manager. If your reason for acting as your own contractor is to save money and you've never contracted a full-sized house before, there is a good chance the house will ultimately cost more than one built by a general contractor.

Financing a log home, like building a conventional home, involves two stages. A construction loan covers the construction period. A permanent mortgage pays off the construction loan over a much longer period of time. Construction loans are simple interest loans requiring the payment of interest only on the outstanding loan balance, while the mortgage payment includes both interest and principle—the amount originally borrowed. Construction loans and permanent mortgages may be obtained separately or together as a construction-permanent loan. As a rule, the combination is far superior for several reasons.

First of all, a construction loan is virtually useless without a permanent mortgage (unless you are certain of having the cash on hand to pay the entire construction loan balance when construction is completed). In fact, most lenders will not issue a construction loan without proof of permanent financing. Second, all loans carry certain fees, usually expressed as "points" or percentages of the loan amount. A $100,000 loan with 2 points will require a payment of 2 percent, or $2,000, at the time the loan is made. In addition, fees for title insurance, title search, appraisal, and mortgage insurance may also be payable when the loan is given. Separate construction and permanent loans will carry two sets of points and fees. Combined construction-permanent financing will require only one set.

If you obtain a combined construction-permanent loan, the construction loan is in effect for the construction period of your log home. During that time, money will be released periodically or "drawn" from the total loan balance to pay for work completed. A "draw schedule" prepared by the bank releases specific percentages of the loan amount at specific stages of completion. For example, completion of the foundation may allow you to draw 15 percent of the loan proceeds. During construction, you will pay interest only (at a slightly higher rate than your permanent mortgage rate) on the amount of the loan that you have actually used. Interest payments are due monthly. When construction is completed, your loan "converts," and the construction loan is paid off by the permanent mortgage. You then start paying standard mortgage payments (principle and interest).

The process for applying for financing can seem overwhelming. Actually, the procedure is fairly simple if you follow an organized approach.

The first step in obtaining financing should be one of the first steps you take toward your log home. Before you have finalized your floorplan, chosen a manufacturer, or obtained a builder, you should know exactly how much money is available for your log home project. Determine this by having yourself prequalified by one or more lending institutions, as discussed earlier. (Shopping for financing is wise, because lenders compete like any other business; a ½ percent savings on a $100,000 loan can mean a lot over the life of a 30-year mortgage.)

Add to the loan amount the amount of cash you have now or will have available for a down payment and you have the maximum amount you can spend on your log home. For your own security, set your budget below the maximum amount you can spend. This margin of error will be helpful when you confront the First Law of Home Construction: Everything costs more than you expect.

For example, suppose you earn $60,000 a year (combined family income), or $5,000 per month. You currently own a home valued at $150,000 with a mortgage balance of $100,000 and monthly payments of $1,000. You have monthly payments of $2,200 including mortgage, automobile, and installment (credit card) debt. You have a savings account containing $12,000 that you plan to use toward your log home. How much are you qualified to borrow, and how much can you spend on your complete log home project (including land)?

Lenders follow a general rule that your mortgage payment cannot exceed 28 percent of your total monthly income. In this example, your maximum mortgage amount would be 28 percent of $5,000, or $1,400. Also, with your installment debt added on, your combined monthly payments on all debt cannot exceed 36 percent of $5,000, or $1,800. Since you will be selling your current home, your combined debt will be $2,200 − $1,000 (old home) + $1,400 (maximum mortgage of new log home) = $2,600. This exceeds your allowable installment debt, so you will have to find a way to reduce your total monthly payments by $800 ($2,600 − $1,800). Assuming you can do that without reducing your mortgage payment, how much can you borrow?

If the lender is offering fixed-rate 30-year mortgages at 9 percent with 2 points, you can borrow $174,000. To this you can add your $12,000 savings and any cash you receive on the sale of your current home, say $25,000 ($150,000 selling price minus $120,000 mortgage balance minus $5,000 realtor's commission and other costs of sale), to give you a total of $211,000. This must cover all of the costs associated with your log home project—land; sitework such as well, road, and septic system; log kit; and construction.

Your lender will specify that the amount of the loan must not exceed a certain percentage of the total project cost or appraised value (whichever is lower). If the lender in our example will lend 90 percent of the cost or appraised value, your project must appraise for $193,000 (90% of $193,000 = $174,000) or more. If the cost of your project is greater than its appraised value, you must make up the difference.

Now that you know what your project limitations are, what about your cash requirements? Let's assume you purchase a kit for $45,000 and establish a budget of $120,000 to purchase land and complete the sitework and house. You have several cash outlays immediately. Your land purchase will require a down payment. Your log kit will require a down payment of 5-10 percent, usually. Your financing carries loan fees of 2 percent (remember those points) payable at the time you settle on (finalize) your loan. In our example this amounts to 2 percent of $174,000, or $3,480. In addition, there will be periodic construction loan interest payments due as you use or draw down the amount of money in your loan account. These payments will be small at first, but they will grow as construction progresses.

Referring to our example: Assume that the construction portion of the loan carries an interest rate of 12 percent annually, or 1 percent monthly, on the outstanding balance. When your loan settles, the bank will use some of the loan proceeds to pay off any outstanding balance on your land. If your land cost $30,000 and you owe $20,000 when your construction loan takes effect, $20,000 will be taken immediately to pay off your land. If you complete the foundation within the same month at a cost of $10,000, your construction loan balance will be $30,000 ($20,000 + $10,000), and your monthly payment will be 1 percent of that, or $300. Next month your kit is delivered ($45,000 − $5,000 down payment) and erected ($25,000) to produce a loan balance of $30,000 + $40,000 + $25,000, or $95,000, with a monthly interest payment due of $950. If you use the full loan amount, your final interest payment could be approximately 1 percent of $174,000, or $1,740.

One way log home buyers get into trouble is by underestimating their cash needs while the house is being constructed. Technically, your loan is to construct the house. It is not meant to include interest payments. Not infrequently, home owners will use proceeds from the loan to make interest payments, which creates a debt that floats along until the house is completed, the full loan amount is used, and $10,000 of unpaid bills to subcontractors and materials suppliers is outstanding. Don't fall into this trap. Account for your interest payments before beginning construction and monitor your budget carefully.

How is a loan application actually made? It's very similar to prequalifying. In addition to the financial paperwork you used to prequalify, you will need a set of blueprints for your proposed log home and copies of your log home kit contract, land contract, and building contract or budget sheet with attached subcontracts. Then you will wait usually two to six weeks. Your paperwork will be lost twice, the appraiser will go on vacation for a month, and someone will spill coffee on your budget sheet making half of it unreadable. Welcome to the wonderful world of home building!

General Contracting:
Where to Begin

FINDING AND MANAGING SUBCONTRACTORS

If there is a single aspect of general contracting your own log home that can make the difference between pleasant adventure and ulcer-inducing nightmare, it is the quality of your subcontractors. A good team of subs can make your log home building experience enjoyable and satisfying, giving you a sense of pride in your home that can't be measured in dollars. Poor selection of subs or one or two "bad apples" in a collection that is otherwise satisfactory, can make you miserable at the very least. At most they can add hundreds and sometimes thousands of dollars to the cost of your project.

Finding good subs generally requires some effort. That effort combines some basic detective work with an ability to judge people. Throughout the process keep in mind that a sense of craftsmanship is only one characteristic of a good subcontractor. Subcontractors who do excellent work are of little value if they can't seem to find the job site with any regularity; likewise, subs who do excellent work only while being closely watched (unless you have the time to watch them and a good understanding of the work you are watching).

Books and magazines sometimes advise the owner-contractor to find and evaluate subs by visiting job sites. Unless you are experienced in a sub's area of expertise, however, it may be difficult to determine the quality of the work you are seeing. I suggest using a referral system to evaluate subcontractor potential.

Sources of referrals are numerous. The most obvious is your log home sales representative. The representative should have a good idea of who in your area is capable and qualified. If the rep can't specify local subs, ask about a sub in another area who is willing to travel. The best referral is from another home owner, preferably someone who also acted as his or her own general contractor. Your log home sales rep should be able to help identify some of these people. In any event, trace your referral to an actual home owner if at all possible. Home owners can give you the best feedback, and they represent an uninvolved party.

Questions to ask when checking a sub's credentials include: Are you satisfied with the quality of the work? Did the work start promptly as scheduled? Was the sub on the job site all day? (Sometimes a sub will leave workers on a job site when there is another job to bid or when running several crews. This may be acceptable for certain trades *if* it is understood at the time the bid is prepared. Be sure of the ability of the workers to function on their own and handle decision making before agreeing to a situation in which the subcontractor will not always be present.) Were the sub and his workers easy to get along with? How were changes received? Did the sub insist on charging extra for even minor deviations from the agreed-upon work? Did the sub volunteer ideas for improving aspects of the project or getting something done more efficiently or economically? Was the job site kept in good condition? Trash picked up? Waste from the work removed or kept neatly out of the way? Did the workday start and stop on time? Did the workers seem motivated and concerned about doing a good job? Did the members of the crew seem to get along?

When getting a referral, try to get answers to as many of these questions as possible. Try for confirmation from a second and even third referral source. Then, if possible, visit a job site where the sub is working. The questions that I try to answer by visiting a job site concern the attitude, efficiency, and cleanliness of a subcontractor. I recently read in a newspaper article an interview with a government official charged with investigating job discrimination. This official claimed that an attitude problem did not justify firing a worker. Baloney! A poor attitude affects the morale of a crew; it leads to shoddy work, mistakes, incomplete work,

work getting off schedule, and more. I'll remove subs from a job site for a bad attitude no matter how highly recommended they come. You can't afford to pay someone to jeopardize your project.

Sometimes you can verify a sub's qualifications or get an indirect recommendation by speaking with your county building inspector. I've seen other owner-contractor books recommend calling the inspections department for referrals, but where I come from this is not done. Inspectors open themselves up for all sorts of potential problems by offering recommendations of subcontractors. Their job is to be impartial in their evaluations. What you can do is follow up a recommendation by phoning the building inspector and asking whether the sub in question has a satisfactory track record with the inspections department.

Finally, when you approach subs about working for you, present them with a written "scope of work." Describe exactly what you expect of them. This actually will become the basis for their bid. Pay attention to how they respond. You should receive a written proposal on a form that contains the company name, address, phone number, and any licenses or permit numbers required to do work in that specialty. If your sub has no phone or mailing address, beware.

SCHEDULING

Scheduling is part science and part art. Some people would even say that it's part magic. The variables on a job site (weather, availability of materials, subcontractor reliability) make scheduling a process of sophisticated guessing at best. But a sophisticated guess is better than none at all.

The first step in scheduling is to obtain time estimates from subs for completing their respective tasks. Find out how much advance notification or lead time they will need in order to get to your job site when you need them. Note these figures on your subcontract forms, and it doesn't hurt to have the subcontractor initial them.

The next step is laying out a rough schedule of the various tasks. A sample schedule is shown in Appendix XI. Use it as a guideline, developing your own schedule to suit your particular circumstances. If you are unsure about scheduling details, you may want to read through the section "Building Sequence" on page 78.

Once you have the construction sequence arranged, add lead times and any special considerations such as building and bank inspections. Note lead times for both subcontractors and suppliers. For example, if you will be ordering materials locally for your subfloor system, note on your schedule sheet that you should order subfloor materials several days in advance of the carpenters' arrival. That way you won't have to face a carpenter demanding payment for a day he couldn't work because there were no materials for him to work with.

Note that often several construction steps can be occurring simultaneously. It's to your benefit to take advantage of these opportunities. They save you time and interest money. Be forewarned, though, that many subs get testy about having to share a job site. Ask in advance, preferably while you are scheduling, before you double up subs on a job site. If a sub is reluctant, listen to the explanation and then use your own judgment. Sometimes their reasons are valid, but not always.

As an example, consider your mechanical system rough-ins (plumbing, electric, and HVAC [heating, ventilation, and air conditioning]). Many subcontractors like to have the job site to themselves. One reason for this is that it gives them the freedom to choose where they place their work. However, if that's the case, the duct workers may put the ductwork where they choose and leave the site; then the plumbers come along and fume because there's a duct where they were going to run a pipe. Now the plumbers have to spend time refiguring a pipe run and maybe even going after additional materials. At the worst, somebody may actually have to remove and redo some work because there isn't any other place for the other sub's work to go. If that happens, you will probably get to mediate a dispute over who pays for time and materials to redo the work.

A simple sequence to follow in this example, assuming you are using an HVAC system that requires separate ductwork (baseboard electric heat is usually handled by the electrician) is to schedule the HVAC rough-in and the plumber together, giving priority to HVAC because ductwork is the least flexible of the mechanical systems. With the HVAC and plumber there at the same time, they can discuss where ducts and pipes need to be and arrange a reasonable settlement. The plumber's time is going to be spent mostly in specific areas of the house (kitchen and baths), so the two subs shouldn't be in each other's way. Schedule the electrician perhaps a day or two later than the other mechanical subs, preferably while they are finishing their work. Although the electrician can deal more easily with obstacles because of the flexibility of wire, should a conflict arise, the other subcontractors will be there to help arrive at a solution.

Similar situations can arise between painters, trim carpenters, floor finishers and fireplace masons. When arranging your schedule, ask subs what work must be completed before they can begin work. Ask if other work you would like to schedule at the same time would interfere. Floor finishers and painters generally require the house interior to themselves. Trim carpenters and fireplace masons often work with other subs.

You will note, as you develop your schedule, that certain construction steps depend on prior completion of another step. Others do not. For example, kitchen cabinets can be installed almost any time after painting is completed. But, framing must be completed (and often inspected) before the mechanical subcontractors can start roughing in their systems. Professional builders often schedule using a "critical path" approach. They lay out the schedule according to the steps that must be done in sequence and then fill in non-critical areas to complete the schedule. You may want to consider this approach. The same schedule in Appendix XI follows the critical path approach.

When preparing your schedule, it is important to note inspection requirements of local building authorities. Before finalizing your schedule, call them and ask for a list of required inspections and at what stage of completion. If there isn't a list available, but there are inspection requirements, take your schedule to the building authority offices and go over it with an inspector. Note inspection times on your schedule and be aware that some building inspectors require twenty-four hours advance notice. For example, if your county requires a footings inspection, make a note on your schedule to contact the inspector to examine the footing trenches before concrete is poured. If the inspector needs advance notice, you may even make the inspection appointment before footings are dug.

It's important to have a thorough understanding of your schedule before construction starts and to monitor and record adjustments throughout the project. You can be assured that, despite your best efforts, there will be some scheduling mishaps—suppliers missing a delivery date or delivering the wrong material, subs not showing up or taking longer than estimated, or weather delays. With a good schedule, you can make adjustments as the situation demands and minimize aggravation and financial impact. If you lose control of your schedule, be prepared for a nightmare at best and a financial disaster at worst.

THE BUILDING SEQUENCE

With your budget established, subs identified, and a rough time frame outlined for your project, it's time to review the overall construction sequence. This should be an extremely detailed analysis of every operation that will go into completing your house. An understanding of the flow of work is vital to establishing a working schedule and making sure that your project stays within budget.

The chart in Appendix XI outlines a detailed construction schedule based roughly on the construction sequence. Review it thoroughly. I have found that the mental exercise of building the house in your mind is a good way to "proof" your building sequence to see that everything falls at the proper place. It can also help you identify bottlenecks and shortcuts that may save you time and money when work starts.

First, as soon as your financing is secured, open a separate checking account devoted exclusively to your log home project. The first check out of this account can be for the deposit on your log home kit. (If you've already written this check, I hope you at least waited until you were sure the financing was in place to carry out your project. Most log home companies require nonrefundable deposits, so if your financing falls through, you may be out a substantial deposit cost.)

Second, obtain a builder's risk insurance policy, available through most comprehensive insurance agents. A verification of this policy will be required by most lenders before they will release any of your construction loan.

Third, apply for your building permit. Depending on your location, this could be a lengthy or a simple process. This will generally require several complete sets of blueprints for your house. As part of your building permit, you also may be required to file a "site plan." This could be as simple as showing the location of the house, well, and septic area on the property, or it may require a complete landscaping plan. In some areas, site plans (or plot plans) must be completed by a licensed surveyor. Find this out as soon as possible (being sure to include it in your budget) and book any necessary survey work as early as possible, preferably as soon as your blueprints are complete.

Fourth, obtain any other permits (such as plumbing or electric) that are necessary but not supplied as part of subcontracts. (Because of the complexities of many permit processes and the sheer frustration of dealing with the permit bureaucracy, my plumbers and electricians are responsible for obtaining the permits for their trades. In some areas, this is a legal requirement.)

Finally, begin contacting your subs, returning signed copies of the subcontract agreements, and filling out your construction schedule.

Site Work

Begin by roughly laying out on your land the location of your house, septic area, and well. Your surveyor may have done this as part of preparing a site plan. If that is the case, there should be surveyor's stakes locating those areas. The house location may be identified with "offsets" — stakes located a specified distance from the front and rear corners of the house. Offsets allow the excavator to work without destroying accurate reference points for the house. Make clear to everyone including yourself that surveyor's stakes are sacred. Survey stakes lost during construction may have to be reset by the surveyor at additional cost to you.

Site clearing can come before or after the house location is marked. If extensive clearing is required, it is best to do it before staking out the house. That way you won't have to stand guard while the clearing crew attempts to work around the layout stakes; it also eliminates the worry of having to pay the surveyor to replace stakes.

With the site cleared, and the house, well, and septic area located, it's time to bring on the well driller. In some areas, location of the well must be approved, and often the well must be inspected and tested, even before a building permit is issued. In any event, don't invest a dime in putting up your foundation until you are certain that your dream home will have a supply of fresh water. Because of the idiosyncrasies of geology, a working, tested well is the only proof you should accept that water is actually available.

Following the well, proceed to the septic field. This gets more of the messy site work out of the way before house construction begins. When putting in the well and septic system, don't let any pipes run into areas where they could be damaged by foundation construction. The pipework to connect the systems to the house should wait until the foundation is in. Note that, unlike the well, the septic field can be installed any time during house construction. Until the house is plumbed and inspected, the septic field is not in the critical path of home construction.

As part of site preparation, you may wish to consider bringing in temporary utilities. These can include electricity, telephone service, water, and portable toilets. Temporary electric service can prevent the cost and aggravation of relying on a generator. Utility companies generally have procedures and instructions for installing temporary service. Temporary water may be available by installing a well pump that can be run from a generator or from the temporary electric service. This can be especially useful if there will be masons working on site erecting block walls and stone fireplaces. A telephone can be indispensable on a job site. It will save its cost many times over, in tracking down lost deliveries and subcontractors, finding sources for urgently needed (but unanticipated) materials, and calling for emergency assistance if needed. The main caution about having a phone on the job site is to control its use rigidly. Construction workers are notoriously transient, and some are delighted to have an opportunity to renew acquaintances in distant locations on someone else's phone bill.

A portable toilet will depend on your location and, more significantly, the location of your nearest neighbor. Some people consider a "port-a-potty" a luxury, but unless you are building in a remote location in a time of favorable weather, I think it's worth the small cost. If nothing more, it will eliminate the vision of errant toilet paper wafting idly across your job site at odd moments.

Foundation

Now it's time to dig the foundation. If the house is going on a slab or piers, this should be a relatively quick process. If you are installing a crawlspace or full basement, it will take longer. Your foundation subcontractor should have forewarned you of any special considerations, such as the potential for encountering rock that would require blasting. When digging a full basement, allow a minimum of 2 feet "overdig" around the walls to allow workers to get around the foundation on the outside to apply waterproofing and drain tile. A 3-foot overdig is generous, and any more is simply creating more work and expense than necessary, besides creating a larger fill area that will restrict the movements of trucks and heavy equipment near the foundation. You do not want to deal, for example, with having a cement truck become mired in the backfill around your foundation, with the potential risk of cracking the wall. Keep the overdig to a realistic minimum.

After the foundation is excavated, footings, or footers, must be dug. These will support your foundation walls and ultimately, along with internal support piers, the weight of the house itself. Requirements for footings are usually stated specifically in county building codes. The depth of the footing must be sufficient to reach below the frost line for the geographic area. Too shallow a footer can buckle and "heave" during prolonged freezing weather. In many areas, building inspectors must examine and approve footing trenches before any concrete can be poured. Fireplace footings should be included along with wall footers.

PHOTO BY JIM COOPER

Footings or footers form the base upon which the foundation will rest. Footings are made either by digging with a backhoe as in the photo or forming trenches with construction lumber into which concrete will be poured. Width and thickness of footers are usually governed by local building codes. In the photograph, the footing is being dug for what will become a walkout basement. To avoid the possibility of freezing soil heaving the footer, it must be dug deep enough for the footer base to rest below the depth of frost penetration. The depth required for such "frost footers" can be found by calling local building authorities or any experienced local excavator.

After pouring the concrete footers and/or piers, the next stage is foundation walls. If the foundation/crawlspace walls will be masonry block, it's time to bring on the masons. If the walls will be poured concrete, it's time to set forms. When foundation walls are complete, they should be waterproofed with drain tile (generally perforated plastic pipe) placed around the perimeter of the foundation. The perforated pipe should be covered with a porous material that will prevent dirt from clogging the perforations when the foundation is backfilled. Special paper called filter or red rosin paper serves this purpose. Foundation drainage should lead to daylight (ground surface) or to a sump hole inside the basement, where a sump pump can carry water away from the foundation. In many areas, the drain tile and waterproofing must pass an inspection before backfilling can take place.

When I was first starting my log home business, I lived in a rental house (not log) built on a flat lot with a block foundation. One night, after several days of prolonged heavy rains, I heard a rumble from the basement. On investigating, I found that a good deal of my yard had come indoors. The entire end block wall of the house had collapsed, carrying tons of mud into the basement. Only by quickly bracing the remaining walls and getting a backhoe (at midnight during a downpour) to dig the backfill away from the rear wall of the house, did we prevent the entire house from collapsing. Never underestimate the importance of proper foundation drainage!

What happens next in your construction process is a matter open to opinion and varies with the type of foundation. If the foundation is block, many builders will proceed to putting on the first-floor subfloor or "deck." This will help brace the block walls as they cure. After the subfloor is in place, the basement floor can be poured. (Some builders wait until much later in construction to pour the basement slab.) Once the slab has hardened, the foundation can be backfilled.

If the foundation walls are poured, the same sequence can be followed as for block. However, because concrete walls have more lateral strength than block, you may choose to pour the basement slab and backfill before putting on the subfloor. I know several builders who backfill as soon as the forms are removed from the poured walls and then pour the slab and continue with the backfill. Discuss the sequence with your foundation contractor to find out what will work best in your circumstance.

My goal in all of the above is to have the basement walls up, basement floor poured, foundation backfilled, and septic system and subfloor in place before beginning any log work. Actually, I prefer to have these things done before the logs even arrive. It is thoroughly exasperating (as well as aggravating, costly, and dangerous) to be carrying logs up gangplanks over a foundation overdig or trying to figure out where to stack logs and building materials amid piles of excavation dirt and topsoil.

Shell Construction

The next step, starting with a completed foundation and no open trenches or excavation piles, is to take delivery of the log kit. If I'm working with a "full" kit that includes millwork, windows, and such, I may bring a storage trailer on site. Storage trailers are not that expensive, and they will provide security from vandals and weather for the delicate portions of the kit. It's virtually impossible to weatherproof and secure a subfloor (while keeping it in a condition that allows construction to proceed), so storage of trim and delicate items in the unfinished basement is not a good idea. Wait until the house has a roof over it and the basement has dried before storing materials there.

The log work and framing is fairly straightforward. Most log home companies have a detailed construction manual that will provide specific information for their product. My construction sequence is as follows: Erect first-floor log walls, followed by the beam system that supports the second floor. Only partition framing necessary to support the second-floor system is done at this time. A temporary second floor is put in place (on two-floor houses, naturally) and log gables are erected. The roof is framed, including any frame gables or dormers, and the roof and dormers are sheathed.

Next the roof is shingled. Now, while part of the crew works outside installing the soffits and fascia, other crew members work inside finishing the log walls (this procedure varies with the style and manufacturer that you use). Interior log finishing includes bleaching interior walls, sanding (staining, if that is desired), and applying a coat of sanding sealer to the log walls.

After the exterior crew has finished its work, the outside wood preservative is applied. Then everyone comes inside to work on interior framing and installing windows and exterior doors. The gutter subcontractors install gutters and downspouts. Windows and exterior doors are trimmed and caulked on the outside, resulting in a weathertight shell that looks complete from the outside. Depending on weather, parking for subs and supply trucks, and the settling of the backfill, I may go ahead and final grade and seed the yard now.

Mechanical Rough-ins

With framing complete and the house closed in, the building inspector usually makes another appearance. With his approval, the mechanical subcontractors go to work. The HVAC sub and the plumber come first, with HVAC getting priority in case both can't start at the same time. As soon as the plumbing is roughed in, the electrician arrives. With careful planning, a week's time sees the mechanical systems roughed in. Then interior wall coverings are installed. These may be drywall or tongue-and-groove woodwork. After wall coverings are installed, I complete painting and varnishing.

Now the house is ready for serious finish work. Ceramic or vinyl floor coverings go down in the kitchen area and baths. Kitchen cabinets and vanities are installed, followed by installation of plumbing fixtures. While this is going on, electricians "switch and plug" the house, installing lighting fixtures and appliances. Once these subs are finished, the trim carpenters go to work. Their tasks include installing hardwood flooring, stairs, and rails; hanging interior doors; and installing interior window trim, baseboards, closet shelves, and hardware (door locksets and doorstops).

Next, I have the hardwood flooring and stairs finished and carpeting installed. (If hardwood floors are involved, I may have the carpenters hold off on installing baseboards and hanging interior doors until after floor finishing.) Last, the painter comes through to retouch any "dings" or marks, and the inspector returns for his last look. (These topics are covered in greater detail in Chapter 9.)

Sweat Equity

"SWEAT EQUITY" CONSISTS of those activities or portions of the construction labor that you undertake yourself, instead of hiring workers. You may do these tasks yourself to hold down the cost of your house, to accommodate your construction schedule, or simply because you like doing a certain kind of work. Whatever the reason, many owner-contractors choose to invest sweat equity in their log home.

Contributing to the construction of your log home can be a financially beneficial undertaking—if you choose your contribution wisely and evaluate it realistically. Unfortunately, not everyone does, and the results can be disastrous. Because many people have only a marginal acquaintance with the world of home building, they can easily delude themselves about both their abilities and the financial significance of their efforts. Almost any banker who has dealt with log home loans can offer horror stories about some poor souls who were going to "do it all themselves." In some cases, banks have used such examples to refuse any consideration of an owner's labor contribution. They want the name of a general contractor or builder on the construction contract and a budget summary that shows labor costs at fair market value—period.

To protect yourself and your log home dream, you need to evaluate any "sweat equity" you are planning in the cold hard light of reality. Unless you have done that job yourself, to the quality standards that you would expect of a professional, don't look at a budget summary and say to yourself, "Harumph! I can do it myself and save all that money. I'm not going to pay that kind of money to do a simple job like that."

Like most log home representatives and builders, I have had experiences with owner sweat equity. I can summarize them like this: Occasionally an owner has performed an outstanding job on a particular task, truly saving money and getting superior work. More often, the owner has saved a little money in return for a job that is below the standards I would consider acceptable for a professional. Then there are more than a few instances in which the owner saved little or no money in return for a job that I would not show anyone.

YOUR SWEAT EQUITY POTENTIAL

Before undertaking any labor contribution, it's a good idea to evaluate your abilities and interests. Even if you can perform a certain task, are you sure that's how you want to spend your personal time? Most owner-contractors have to use weekends, evenings, and vacations to work on their log home projects. Before further committing those precious personal hours, take a moment to ask yourself: Is this how I want to spend my beautiful autumn weekends this year? If the answer is yes, proceed to the next question: Can I do a job that is technically competent, safe, and appealing to look at? Then, on to the final question: How much will it cost me to do the work myself?

When evaluating savings, be realistic. If the cost of having a subcontractor trim your house, for example, is $3,000 for materials and labor, don't make the mistake of saying, "I can save myself about $2,500 by doing it myself. Materials can't cost more than $500." That way lies disaster.

Instead, perform an accurate "take-off" to estimate the actual materials you will need. If you aren't sure how to calculate material requirements, get a good book on that specific topic. Don't proceed if you aren't confident in your estimating ability. Owner-contractors get themselves in trouble ranging from inconvenience to having to sell their house before it's finished, simply because they misjudged the work or materials required to do some particular task they thought would be easy or inexpensive.

Always include an estimate for your time. Even if you are not planning to pay yourself, calculate how long you expect to take to do the work and about how much you would pay yourself. Put these numbers into your budget and schedule to make sure the project remains workable. Don't overlook the effects on construction interest of performing your own labor. Let's say that a subcontractor estimates that your trim will take about two weeks and he will charge $2,000 for materials and labor. You know that you are capable of doing the work, but you can only work on weekends. You calculate material costs of about $700 and estimate that you will need about six week's worth of weekends to do the job.

The final trim is the only thing standing between you and the final inspection that will allow you to convert your interest-only construction note to a permanent mortgage. The extra month you require to do the work means an extra month of construction interest. If that will run $1,000, your potential savings is only $300, not $1,300. Do you want to sacrifice those weekends and wait an extra month to move for $300?

It is becoming more difficult to underestimate sweat equity these days, thanks to lending institutions. Many lenders now require you to include in your budget the fair market value of any labor that your contribute. In the above example, the construction loan and mortgage would be based on a figure that included $2,000 for trim work. The lender's reasoning is that this approach insures enough money budgeted to complete the house even if you, for some reason, do not do the work. (You may go ahead and do the work, paying yourself from the construction loan or simply leaving the money in your construction account to lower the amount of your mortgage when the house is complete, so you can still realize a savings.)

My advice then, for evaluating potential sweat equity contributions is to ask yourself three questions: Am I capable of doing the work to my standards and the requirements of building and safety codes? How much money can I expect to save by doing the work myself? Do I want to do the work myself? Only when you've answered these questions honestly are you in a position to include sweat equity in your log home plans.

What follows is a review of the major areas of construction labor and my own evaluation of their potential for sweat equity contributions. The descriptions are not intended as instructions. For detailed descriptions of actual construction methods, visit your local bookstore or contact one of the references listed under "Resources and Supplies" in Chapter 11 at the rear of the book.

Foundations and Slabs

One customer I had dug his own basement. He bought a bulldozer, had a friend (professional) give him some instruction, and went to work. He wasn't fast, but he did a creditable job. This, however, goes beyond what most people would consider sweat equity. Unless you have heavy equipment in your background or in your blood, you will probably want to leave this to a subcontractor.

You may, however, choose to lay out your own house site for digging the foundation. This isn't difficult; a good book on house-building, even the instruction manual with your log home package, may give you the necessary information. One word of caution: Be sure the foundation is on your property! That's not a joke. More than one person has laid out a foundation and started to dig, only to find that he has ventured onto the neighbor's land or outside the legal "set-backs" of the property. Be certain of your location before you proceed.

Some owners lay out their foundation walls. Here you will need a working knowledge of a builder's level or transit. These two instruments are not identical but have many similarities. A builder's level operates only in a horizontal plane, reading height on a graduated stick. It is used to check excavation and footer depth, wall height, and other vertical measurements. A transit offers the same height measurement capability, plus it can be angled, allowing it to be used for locating points on a line and marking corners (these tasks require more knowledge of geometry and some trigonometry). Another layout tool used by some is a water level. This consists of two cylinders graduated in inches and fractions, joined by a water-filled hose. Water levels are used to compare heights and determine level for points that are separated by long distances (which a builder's level or transit can also do) or a corner (which a level or transit cannot do, at least directly).

Masonry work such as laying block or stone is not beyond the scope of an ardent do-it-yourselfer. However, the capacity to foul up a project is enormous, and unless you have prior experience or are working on a small "weekender" log home, I wouldn't recommend undertaking this aspect of construction. Besides, the speed of most masonry crews usually more than justifies their expense. Save your efforts for other areas.

Poured-wall foundations are also generally left to professionals. First of all, such walls require special forms. Second, the forms require considerable expertise to set properly. Again, on smaller projects, owner-builder publications sometimes describe do-it-yourself methods involving plywood forms (sometimes the plywood is then recycled into subflooring). But this work is generally best left to someone with previous building experience.

Then there are the increasingly popular wood foundations. To me, these seem to offer the best opportunity for do-it-yourself labor. The skills required are primarily those of a carpenter—building studwall sections, plumbing, leveling, and insulating. If nothing else, the weight and type of materials involved make it more likely for an owner to do the work quickly and without excessive physical effort. (One of my concerns about the other types of foundation is that the physical labor required is not the kind of labor familiar to most people.)

Finally, there are precast concrete foundations. These offer a tremendous time savings and so may be worth considering, but they really don't offer much potential for owner input.

Once the foundation walls are in place, a couple of tasks must be completed prior to backfilling. The foundation must be waterproofed and drain tile installed. These tasks are well within the scope of most people. A number of waterproofing compounds can be applied to the exterior of concrete or masonry walls. (Masonry walls should be "parged," that is, coated with portland cement on the exterior surface. In many areas this is a code requirement; it is usually part of the mason's subcontract. Parging is not the same thing as waterproofing a foundation.) The most basic waterproofing is a black tar applied with a brush or roller. Some dry-mix powders are used, also, and are applied with brushes, rollers, or trowels.

Laying drain tile is generally not a difficult task. Drain tile is actually a misnomer, because today most drain tile consists of 4-inch perforated black plastic pipe in snap-together sections. This is laid around the base of the foundation with a drainage pipe leading away from the house at a sufficient grade to carry off water. Paper, called filter paper or red rosin paper, is laid atop the pipe to prevent dirt from filling the perforations and being carried into the pipe after the foundation is backfilled. A layer of gravel should be used as a bed for the drain tile, with scoops of gravel to hold the paper in place during backfilling. Ability to use a small tractor or skid loader can come in handy here. Otherwise, plan on shoveling and wheelbarrowing a lot of gravel.

In the final analysis, my advice to my customers is this: In the absence of experience with the construction methods and materials used in the foundation, leave the labor to professionals. Learn to use a builder's level or transit so that you can keep an eye on the work being done. If the law will allow and you feel comfortable with the task, consider staking out the foundation yourself and, if you like exercise, do your own water-proofing and drain tile work.

Building a Subfloor

I've dealt with a number of people who have set their own subfloors. A knowledge of basic carpentry is all that's required. The important skill is the ability to keep things level and square. Setting a subfloor consists usually of setting a beam of steel or doubled- or tripled-dimensional lumber (usually a 2 x 12) into beam pockets in the foundation wall and supporting it along it's length by wood or steel support posts. The success of the subfloor and all construction that rests on it starts with getting that beam level. Then dimensional joists are placed on top of the beam at right angles. These are fastened to a band board around the perimeter of the subfloor. Last, sheets of tongue-and-groove plywood are laid atop the joists and nailed and glued into position. A good book on basic carpentry can provide details.

If you are going to do your own subfloor, take the time to learn how to level the sill and square the subfloor. Sometimes foundation walls may not be level or square. If the sill is not leveled, the subfloor will be out of level. If the foundation is out of square, and you don't take corrective steps at the subfloor stage, you will be forever recalculating dimensions on your blueprints.

Raising Logs

Recently, I met with a log home customer who lives in a log home that he built himself. He told me it was an effort to get his wife to visit our display home. After lengthy negotiations she agreed to come and look, but only if he promised he would not get involved in building. She said she couldn't go through that again.

Actually, I've talked with a number of people over the years who have built their own log homes. Many say they wouldn't trade the experience for anything—they also wouldn't repeat the experience for anything. They love log homes and have a great deal of respect for hired labor.

On simpler houses, log work is not that difficult. It is, however, physically demanding and tedious. On larger or complex homes, the joinery required becomes more critical to the structural integrity of the house. A house with log bays and many corners offers a real challenge, and the potential for problems is serious. When I first opened my business, I looked for builders. One particular builder impressed me with the quality of his workmanship. I asked if he had ever built a log home. It turned out he had and, thank you very much, but he would skip the opportunity to repeat the experience. It seems that his company had been asked to straighten out an owner-builder project. The owner had started his own log work and, by the time the walls were complete, the corners had the look of circular staircases. The owner had checked each course for square with a framing square, but he wasn't aware that just because the logs of one course were square to each other didn't mean that they were correctly aligned with the logs beneath them. As the walls went up they started to wind. So the owner turned to professional help. After being unable to rack the walls back into position with a come-along (similar to a winch), the crew was forced to dismantle the log walls for about half their height. This experience turned my builder friend against future ventures in log home building.

Had the owner known what to look for, his walls never would have wandered in the first place. His construction manual simply did not contain enough detail. Before you say, "I won't have to worry about that because I will buy a log home kit that is already precut," let me say this: The above parable occurred on a precut kit. Precutting is no substitute for construction knowledge.

To elaborate further: I represent a company that does not precut kits. Quite often I bid against precut packages. Usually, there is a substantial price difference — far beyond the actual cost of the labor involved in cutting logs to length. To potential customers who are planning to build their own house I say: If you are not comfortable using a tape measure and making a simple square cut, I'm not sure I would recommend doing your own carpentry. What is the difference between cutting a log and cutting a 2 x 4? It's the size of the saw, that's all. Otherwise, the same skills are required. So you are paying a lot of money to avoid one of the simplest operations in the log building process.

Beams and Trusses

Beams and trusses are generally included with the log construction. The same tools and skills apply. I have worked with people who erected their own log walls and then brought in a professional crew to frame the roof system, particularly when heavy beams were involved. Raising heavy timbers safely and efficiently requires a skill apart from carpentry. Even many framing carpenters lack the knowledge to do it well. If your log home has lots of beams or an elaborate roof system, you may want to consider leaving this to a pro even if you raise your own log walls.

Setting conventional trusses, on the other hand, is not exceptionally demanding. It's actually gratifying to see how fast a conventional truss roof can be readied for sheathing and that exciting day when rain or snow can no longer impede your progress. If you've never worked with trusses, you might want to visit a job-site and watch. Bracing, squaring, and spacing are especially important. Conventional trusses should arrive with working drawings and instructions or construction details. You can learn more by calling a local lumberyard or truss manufacturer. (Truss manufacturers are located everywhere, either as part of lumberyard operations or as separate businesses.)

Roofing

By roofing, I mean sheathing, papering, and shingling. Roofing is not especially difficult as long as you are not afraid of heights and have a fairly good sense of balance. The most difficult task is getting the plywood sheathing onto the roof and into position to nail. Experience with a nail gun helps here. Otherwise, be prepared for a lot of hammer swinging.

Shingling is much easier than getting the shingles onto the roof. If shingles are a part of your log home kit, you will want to rent an elevator or some kind of lift to save your back. Once they are on the roof, shingling is fairly straightforward. This is an area, however, where neatness counts, because misaligned shingles or crooked lines are going to be painfully noticeable. Another consideration is the complexity of the roof; shingling hips and valleys requires more skill than straight runs.

Along with the shingles, areas of potential leakage must be flashed and sealed. Remember that along with the cost of a professional roofer comes your license to pick up the phone and say, "Come fix this leak." That should be worth something. Try watching some roofers at work, or check out some do-it-yourself books or videos before deciding to undertake this task.

Interior Partitions

After a professional crew has erected the log walls and "dried-in" the house, some homeowners take over and complete the work once the house is under roof, finishing and framing and directing the mechanical subcontractors. There are a number of good books on carpentry and some excellent videos that show the intricacies of framing. It's not particularly difficult but will proceed much faster if you've had some experience.

Electrical Systems

Many areas require an electrician's license to become involved in residential wiring. If the laws in your area allow it and if you have a mechanical aptitude, wiring a house can be fun. You will need some special tools to undertake your own electrical work. Generally these are not expensive or complicated; their cost, however, may be a factor in deciding whether to do it yourself or subcontract. Review books or videos and, if possible, watch an electrician at work.

Plumbing Systems

As with electrical work, many areas require licenses. Beyond that, owners that I've known have seemed more comfortable with wiring than with plumbing. The advent of PVC pipe, however, has made plumbing much easier because it has replaced soldering with gluing. Approach the decision of whether to do your own plumbing work the same as with electrical work.

Heating and AC Systems

Few people seem to get involved in installing heating and air conditioning systems. I don't blame them. The technology is complex; there are some special tool and knowledge requirements; and working with sheet metal is not particularly pleasant. Besides, there is not a great cost savings to be gained (in my experience, anyway).

Kitchens and Baths

There are several areas of kitchen and bath finishing that you may wish to consider. The savings will not be enormous, but they can add up. First are floor coverings. Application of sheet vinyl and many kinds of ceramic tile is relatively easy for someone with a perfectionist streak (or an indifference to precision). Numerous publications and videos explain details, and specialized equipment is minimal and inexpensive. If you've ever replaced a floor covering in an existing home, you'll be amazed at how much easier the work proceeds in new construction.

You may also elect to set your own cabinets and vanities. This work is more exacting, but well within the scope of most handy people. Another potential advantage for undertaking this kind of work is that its position in the construction schedule is more flexible. You won't have to worry about interfering with the progress of the house if you must delay a week before installing cabinetry.

There's nothing mystical about hanging medicine cabinets and mirrors, either. (Be sure to anticipate recessed medicine cabinets during framing.) You may also want to consider ceramic tile counter tops for kitchen cabinets and vanities. These can be striking, and they are not extremely difficult to install. The material cost is low, and many people enjoy the creative challenge of tiling.

Trim Carpentry

I recently visited with a banker who said he will no longer consider owner-builders when making construction loans, because of their generally lousy workmanship and the interminable amount of time that it took them to complete a project. "What if they only want to do the trim work?" I offered. "That's probably the worst place to turn an amateur loose," was his immediate reply. "More than once I've had to replace home owner trimwork when the owner became unable to complete the project for one reason or another."

I have noticed in a number of houses that much of the self-styled craftsmanship in trimming the home is not up to professional standards. Some people say I'm too picky, but my personal taste is for craftsmanlike work on the trim.

That said, I do believe that a reasonably handy person with good tools can do a creditable job of trimwork. There are some cost savings to be gained, and generally the work schedule can be flexible. But be careful. Many log home owners move into an untrimmed house and three years later are pointing to simple things like baseboards that remain incomplete. So many, in fact, that it's almost a standing laugh among log home sales people. Because there's no urgent need to complete trim work (usually, it's not necessary for an occupancy permit) the innate human tendency to procrastinate takes charge. I've been in homes where the owners lived happily on stained and varnished subflooring without a stick of trim in sight. "We moved in and decided, what's the rush?" goes the rationale.

If you intend to do your own trim work, review books and videos and decide exactly how you want your house trimmed. Study photos of other log homes. (Recognize, too, that the need or desire for some trimwork may not become evident until the house is nearly complete.) With a rough idea of the trimwork requirements, get a bid from a professional trim carpenter and an estimate of the time required. Then ask yourself, "How fast am I likely to get these things completed if I move in right as the beautiful fall (or spring) weather arrives? Do I want to spend weekends cursing corner molding cut 1/16 inch too short and miters that don't come together?" Be realistic.

Painting

Many log home owners elect to do their own painting, staining, and varnishing. Most people are acquainted with the techniques and may already have the tools. The potential for savings is substantial since labor cost covers the bulk of a subcontract figure.

The best time to paint is before doors are hung, floor coverings installed, and any other trim put in place. This means that you can slow down your project, should you try to fit these activities around your regular schedule. Be sure you are available to do the work when it is best done. Otherwise, you run the risks of losing subs, incurring additional costs (including construction loan interest), and having to work around components that should have been left alone until after painting and staining were completed.

Whenever stained trimwork is involved, I suggest that you tackle it as soon as possible. Simply stain, varnish, and store the trim until the house is ready for it. That way you won't have a carpenter appear one day saying, "I'm ready to hang doors," when you have bare doors that will take a week to finish (remember there is drying time involved in this kind of finish work.) Your only alternative to holding up the carpenter is to have him hang the bare doors. Then you must sand and finish them in a very uncomfortable vertical position, instead of being able to lay them horizontally. You will also have to mask the areas around the doors and cope with the inevitable varnish runs that occur when varnish is applied to vertical surfaces. In other words, you will have more than doubled the amount of work you have to do, substantially increased the time needed, and reduced the quality of the workmanship that is possible. Worst of all, through the whole ordeal you will be perfectly aware that all of your frustration could have been avoided had you simply started the work sooner.

Before undertaking painting, staining, or varnishing, I suggest reviewing some of the excellent do-it-yourself videos and books. Even if you've done this kind of work before, unless you've watched or worked with pros, you are probably not aware of the many time-saving tips, shortcuts, and "trade secrets" that can make these tasks quicker and easier, with higher-quality results.

Finishing Touches

The multitudinous details that make a log home "finished" range from hanging door knockers and mailboxes to installing doorstops. Some of them will be covered in subcontracts; others will, through simple over-sight, fall to you. You may be admiring your new kitchen cabinets as the plumber crawls out from hooking up the dishwasher. He will raise the question of who levels it and hooks up the electricity. Oops. You can call the electrician, slip the plumber an extra twenty (if he's willing and you trust him), or you can take the plunge yourself. The owner's manual should light the way.

Sweat equity can offer cash savings and a measure of satisfaction that goes beyond that of supervision. The key is to understand thoroughly your motivation for getting involved in the physical labor and analyze carefully the financial effects of your involvement. As long as you don't delude yourself about your abilities and the amount of money you may save, you may find the fruit of your labor can be uncommonly sweet.

9

Construction Process

BASIC SKILLS AND TOOLS

Even if you do not plan to contribute physical labor to the construction of your log home, there are several tools and skills that I recommend you obtain and master. These include a 30-foot steel tape measure, two 100-foot tape measures, steel framing square, high-quality 4-foot level, builder's level (rented), water level (optional), and calculator. (A construction calculator is a good investment. It works in feet and inches directly rather than requiring dimensions to be converted to decimal equivalents. These calculators are advertised in carpentry magazines and usually cost less than $100.) These items are used to check the progress and the quality of the workmanship of your house. By quality I am referring to four basic conditions: square, level, plumb, and dimensional accuracy.

Some people might suggest that monitoring these items is unnecessary because subcontracts state that all work must meet acceptable standards. They feel that, should an error be discovered, it is straightforward matter to tell the appropriate subcontractor to fix it. But what if the error isn't discovered until late in the construction process? Then, even if the subcontractor eats the cost of tearing out and repeating his work and absorbs the cost of replacing any work of other subcontractors that are affected, you, the home owner, must still suffer the loss of time. It's much easier to keep an eye on work quality and catch mistakes before they become the basis for more errors. Even good craftspeople can misread a tape or level.

Framing Square

"Square" refers usually to walls, building components (such as foundations), and framing members being aligned at perfect right angles. Corners, doorways, and window openings can be checked for squareness using a framing square. Before using a framing square, be sure that it has not been dropped or bent out of alignment. Lay the square on a flat surface along a straight edge or against a wall. Draw a sharp pencil line along the perpendicular leg of the square. Now flip the square end for end and lay it back along the same edge or wall. The pencil line should still match exactly the perpendicular leg of the square. If there is any deviation, either replace the square or correct it using instructions supplied with it.

Understand what you are checking when using a framing square. For example, squareness of a corner pair of logs does not guarantee squareness of the wall. Checking the squareness of foundations, subfloors, and rooms requires a little mathematics. A foundation, subfloor, or other large area is checked by comparing measurements of the diagonals (distance to opposing corners). The area being considered must have four corners and be a rectangle or a square. Areas with more than four corners should be broken into rectangles with intermediate measurements. An area is square if the two diagonals have the same measurement. In checking squareness, the

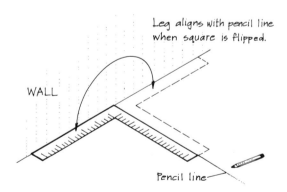

Leg aligns with pencil line when square is flipped.

WALL

Pencil line

TESTING A FRAMING SQUARE

actual diagonal measurement is not important; what matters is simply whether the two diagonals are equal.

Checking dimensional accuracy or laying out rectangles on a large area requires use of the mathematical law that says the sum of the squares of the two sides of a right triangle is equal to the square of the hypotenuse (the diagonal, when the right triangle is made by dividing a rectangle with a line between opposite corners). Consider, for example, laying out a rectangular foundation. You've placed two stakes marking the ends of a 30-foot end wall. Now it's necessary to locate the stakes for the opposite end wall exactly 50 feet away and parallel. You can't simply measure 50 feet from one stake and then measure off 30 feet from that point, because you have no guarantee of forming square corners.

Instead, add the squares of the two sides of the foundation (30 x 30) + (50 x 50) = 3,400. Find the square root of 3,400 (58.3095, or 58′ 3²³/₃₂″). This is the diagonal measurement for the foundation. Using two steel tape measures, one attached to each of the two stakes, measure out 50 feet on one tape and 58′ 3²³/₃₂″ on the other (this number can be rounded to 58′ 3³/₄″). The intersection of these two distances is the point to

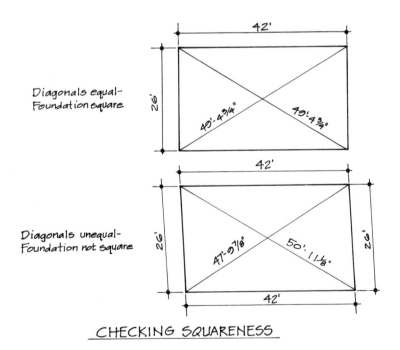

CHECKING SQUARENESS

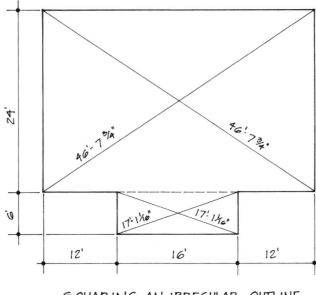

SQUARING AN IRREGULAR OUTLINE

place your third stake. Now repeat the process reversing the measurements on each tape. This will locate your fourth point. As a final check, a measurement taken between the third and fourth stakes should be exactly 30 feet.

Checking squareness is fundamental to quality control on your log house. If the foundation is out of square, the house must either be out of square or (if the error is slight) sit unevenly on the foundation. Interior partition walls that are not square mean more difficulty and time in trimwork, installing floor and wall coverings, and even in furnishing rooms. Take the time to learn this simple skill and follow this rule: If you do the layout, have someone else check its squareness (doing the necessary calculations); if someone else does the layout, verify squareness with your own calculations and measurements.

Levels

You can verify whether horizontal surfaces are level and vertical surfaces plumb using a carpenter's or mason's level. A 4-foot level will handle most situations, although a "torpedo level" or 2-foot level may be necessary for tight places. Before you can evaluate workmanship you must be sure that your level is accurate. These tools take a beating on job sites and can easily become unreliable. One fall or bad bounce in the trunk of a car can mean tilted window openings or walls, or sloped floors.

It's easy to check the accuracy of a level. First, for horizontal (level) accuracy, place the level on a smooth horizontal surface. Note the exact position of the bubble in the window. Now reverse the level and place it in exactly the same location. The bubble should appear in the same position it was in before. Now flip the level so that the top edge becomes the bottom and repeat the procedure. Realize that the surface need not be perfectly level. What is important is that you note the exact position of the bubble in the vial. If the bubble does not remain in the same location in the vial in all situations, the level is off and must be adjusted or replaced. Inexpensive levels are often nonadjustable and difficult to realign. A high-quality level usually offers a means of realignment using a procedure explained by the manufacturer.

To check "plumb" or vertical accuracy, use several feet of string line with a weight attached that allows it to hang free (called a plumb line). Simply hold the level with an edge along the string and note whether the

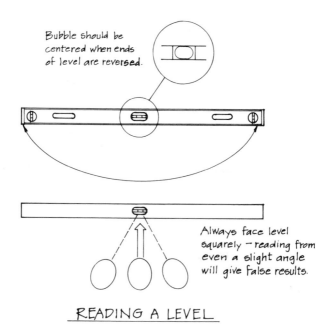

Bubble should be centered when ends of level are reversed.

Always face level squarely — reading from even a slight angle will give false results.

READING A LEVEL

appropriate bubble(s) are centered between markings. Any deviation must be corrected or the level is useless.

When verifying work on your log house, don't rely on a borrowed level or worse, the same level being used to perform the work. Use your own level, storing it in a place where it will not get knocked around or risk being used as a pry bar by a carpenter's assistant. Get in the habit of monitoring level and plumb often as work progresses on your house. Take random readings and point out any areas that seem to need correcting. If your carpenter gets huffy, explain that your checking doesn't mean that you lack faith, but rather that you are simply verifying, like the proofreader of a book, that something hasn't slipped by unnoticed. You can point out that your checks help protect him against other subcontractors claiming, "We didn't do it, that's the way it was framed."

Builder's Level

A builder's level is used to measure elevations. It can be used to check slopes for amount of fall, excavation and footer depths, and levelness of foundation walls. A transit can perform the same tasks plus several more (that require more mathematics than many people want to tackle). These instruments are fairly expensive and their limited usefulness makes it more economical to rent them.

Set up the builder's level according to instructions supplied with the instrument or directions contained in a good general construction text. Using a builder's level is a two-person operation; one person operates the level, while an assistant holds a graduated measuring stick. The level should be set up where there is an unobstructed view of each of the points you want to check (such as house corners). The level location should also allow the elevation stick or "pole" to be read directly. For example, don't set the level on a slope where a line of sight will hit the ground before reaching the point you want to measure. Set the level higher and get a longer elevation stick, if necessary.

The builder's level is used by taking comparative readings of the elevation pole. The pole is rested on the point to be measured and held perfectly plumb (using a level to maintain perfect plumb will produce greater accuracy). The operator sights through the level and notes the elevation indicated by the crosshairs. The pole is moved to the next point and another reading is taken. By comparing the two readings, the relationship of the points can be determined. (One point high, low, or level in relation to the other.) By measuring against a reference point, the depth of excavations and footers can be determined. Even if you don't plan to shoot your own readings for your excavation and foundation, it's good to know how the process works, so you can understand your foundation contractor or excavator should any elevation questions arise.

As an example, let's say you've set the level to monitor excavation. You have a flat lot and you know that your excavation needs to be 7' 10" deep. Before digging begins, a measurement in the excavation area reads 4' 10" — the height of the builder's level above the ground. Adding 4' 10" to 7' 10" will give a total reading of 12' 8". This should be the reading at your finished excavation depth. As excavation proceeds, you can shoot measurements into the excavated area periodically. When the elevation pole reads 12' 8", you've reached the required depth. As you approach the desired depth, it's best to get readings from several points in the excavation. That way you can make sure the bottom of the excavation is level. (If one part of the excavation is deeper than another, someone is going to be paying for extra concrete or stone fill to produce a level basement floor.)

A good use of the builder's level (and excellent practice) for owner-contractors is checking the slope of the house site to get an idea of how the finished house will look. This can be fun, and there isn't the pressure of an excavator or foundation contractor breathing down your neck. Pack a picnic lunch and a set of your house elevations (drawn to scale), rent a builder's level, and head for your site. If possible, set the level to allow you to read all corners of your foundation from one position. Start with what will be the deepest corner of your excavation and take an elevation reading, noting it on the appropriate corner in your elevation drawings. Now proceed around the foundation, recording each measurement on the elevation drawings until you have a reading for every corner.

Elevations are often drawn with the basement shown as dashed lines. If no basement is shown, use a ruler or architect's scale to pencil in an outline of your foundation. Note the high corner (the corner with the smallest measurement) and mark the corner at what will be ground level. Use a ruler or scale to locate the point, remembering building codes generally require at least 8 inches of foundation above ground, and 12 inches isn't a bad idea. Proceed to the other corners, calculating their height in relation to the first corner and using the ruler or scale to mark the appropriate height on each corner. By connecting the marks you can see how much foundation will be exposed on all sides. This can be extremely useful for planning decks and positioning the house for basement walkouts.

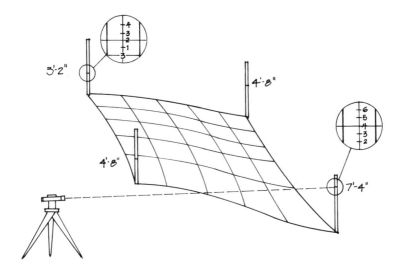

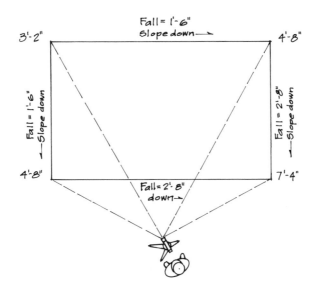

USING A BUILDERS LEVEL

Tape Measure

Tape measures are useful for checking dimension accuracy. Get in the habit of carrying a tape whenever you visit your log home site. Check window and door locations, interior wall positions and lengths, and window and door openings. I actually start window and door measurement checks in the blueprint stage. I check the opening dimensions shown on the schedule page of the blueprints against the dimensions shown in the blueprints; then I check these dimensions against the unit dimensions given in the window manufacturer's catalog. Then I check the openings themselves when they are in place. I also check the catalog numbers of the actual unit against those given in the window schedule. All of this prevents waiting for a new window unit because the wrong size was shipped, or reframing a window opening because a dimension was recorded wrong on the blueprints or in the schedule.

Water Level

Water levels are handy for checking level along the top of a foundation wall or deck. They come in handy for retaining walls and places where points are separated by obstacles that would make use of a builder's level difficult (on opposite sides of a wall, for example). A water level works on the principle that gravity will equalize the height of the ends of a column

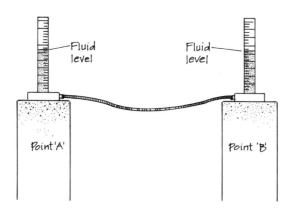

WATER LEVEL

of water contained in a flexible tube. A water level is an inexpensive and simple instrument, consisting of a flexible tube with graduated columns at either end. The tube is filled with water, usually containing food coloring or antifreeze, and the columns are clear to allow you to see the height of the fluid inside. The columns are placed on two points to be measured and the fluid allowed to stabilize. Simply reading the height of the fluid column in each tube tells you how the two points compare. An excellent use of a water level is along the top of a foundation wall where readings will allow carpenters to shim the sill plate upon which the subfloor will rest. Comparisons with foundation walls and the top of the first-floor support beam can insure a level subfloor. An additional advantage of a water level is that it can be operated by one person.

SITE PREPARATION

Job-site Plan

Site preparation begins with a job-site plan based on a surveyor's plat or site plan, or a sketch that you prepare. It should indicate the location of the house, well, septic system, road, utility lines, material storage areas — particularly where logs will be stored — material storage trailer, portable toilet, and topsoil and excavation soil mounds. It's best to develop this plan with the help of your subcontractors. They can point out any special considerations that particular trades might involve. For example, your excavator can warn you if you are preparing to locate the storage trailer where it would be best to store topsoil and earth from the excavation. The plan should also indicate any natural features you want to protect and any sediment control measures you plan (or are required) to install to prevent runoff of earth into waterways.

Clearing

Once you have a plan sketched, you can mark trees that must be removed, using plastic ribbon available at most hardware or building supply stores. Mark the trees close to eye level with a couple of wraps that will be easy to see from a distance. Almost everyone wants to save as many trees on a wooded lot as possible, including trees that may sit close to their future house. This is noble but, as one who has spent several thousand dollars over the years in vain efforts to save trees that simply couldn't be saved, let me offer some advice. Trying to save any tree closer to your excavation

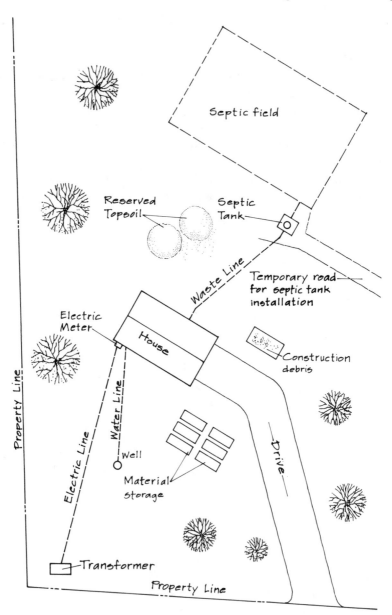

Septic field

Reserved Topsoil

Septic Tank

Waste Line

Temporary road for septic tank installation

Electric Meter

House

Construction debris

Property Line

Electric Line

Water Line

Well

Drive

Material storage

Transformer

Property Line

JOB SITE PLANNING

than the width of the crown of the tree will probably be futile and may be costly. Felling a towering oak that stands right next to your house a year or two after the house is built will require a lift truck or crane and skilled tree-removal experts with hefty insurance policies. They will have to drive heavy equipment across your landscaped yard and park close enough to the house to risk damaging the foundation. They will leave enormous ruts, sawdust, bark, and (unless you really like spending money) an enormous pile of firewood that should be removed immediately from the proximity of your house before it becomes a staging area for wood-eating insects.

It's best to deal with such situations before the house is built. Make sure that you allow a "bulldozer zone," removing enough trees to allow a loader or bulldozer to backfill the house without having to negotiate a thicket of tree trunks; 15–25 feet is a minimum here. Make your apologies to the tree, flag it for removal, and put a tree dedicated to its memory in your landscape plan.

Tree removal is usually straightforward on a bare lot. If there are outbuildings, stored materials, and equipment in the way, it becomes more difficult and costly. Many people I've worked with have done their own tree removal using chainsaws. When cutting down trees, I generally leave 3–4 feet of trunk with the stump. This way the loader operator can use his machine to push the stump over and break it loose from the ground. Stump removal is then easier and faster than cutting the tree right at ground level, which forces the loader operator to dig it out completely. Downed trees are best cut into firewood lengths and stored out of the way of construction activities. If you do not plan to keep the firewood or will have more than you want to keep, you may be able to trade wood-cutting services for firewood. A classified ad can put you in contact with someone willing to cut down trees in return for firewood. In my area, carpenter ants often nest in dead or dying trees. Destroying their nest sends them searching for new accommodations. Stored firewood should be far enough away from the house site (at least 50 feet) to discourage insect migrations.

Good lumber trees may have better uses than firewood. On one job, I had to remove several magnificent red oaks. I found a woodcutter with a portable sawmill who cut the oak lumber into 1-inch-thick planks. These I took to a mill where the lumber was kiln dried and milled into hardwood flooring. The lumber eventually became flooring and trim for Laurel Lodge. I estimate that my final cost was slightly less than half what I would have paid had I purchased kiln-dried red oak from a lumber yard.

Stump Removal

Rotting stumps are havens for insects and invitations to insect problems in your log home. Furthermore, as the stumps decay, a hole develops that can play havoc with lawnmowers, bicycles, and inattentive walkers. Gather up stumps and burn them, if permissible, or have them hauled away. Some people bury stumps and other debris near their job site. The result can be an eyesore when the ground starts to sink as the stumps decay.

Rock

Surface rock can often be cleared using a bulldozer, loader, or backhoe. Rock outcrops or sunken rock seams often require heavier equipment or dynamite. Some log home owners use local rock to build retaining walls, walks, fences, foundations, or fireplaces. The fireplaces at Laurel Lodge are made from rock found on the ground surface within several hundred feet of the house. Sorting and stacking rock is tedious and back-breaking labor. If you plan to do this yourself, allow plenty of time in your schedule and lay in a supply of muscle relaxants. It's not hard to find laborers who will do such work for a reasonable wage.

Because it's often impossible to tell from the ground surface what lies underneath, most excavation subcontracts contain a "rock clause" that states that rock removal is not included in the quoted contract figure. When rock is encountered, the excavator may try to work it out with his loader or a track hoe (large backhoe on tracks similar to a bulldozer). This may be sufficient if the rock is laced with fractures or is arranged in sheets like shale. If the rock is solid, a hoe ram may be called in. This is a backhoe with a large jackhammer on the rear arm. Large masses of hard, unbreakable rock require the use of dynamite. The cost of these solutions generally increases with the size and number of machines required. Dynamite is often the most expensive (but perhaps the only) alternative. If it comes down to a decision among solutions, remember that it may be less expensive simply to blast and clean up than to get a lot of heavy equipment pounding away for days.

Drainage

Even flat lots present potential drainage problems, making it important to consider where water goes when it hits the ground before your project starts and after construction is underway. You need to consider both the water itself and soil runoff. In many areas, building permits require placement of silt fences to protect waterways and adjoining property. Courtesy requires that you consider potential runoff onto neighbors' land even if there are no legal requirements.

Silt fences can be obtained from any good building supply outlet. The fences usually consist of cloth or plastic fabric stapled to wooden stakes. The lower section of fabric is usually not attached to the stakes, creating a flap that should be buried slightly to prevent water from running underneath the fence. It's best to ask your local building inspector how to install a silt fence properly. An improperly installed fence is worthless at best and may actually serve to make things worse. In some areas, home builders can be fined for improperly installed silt fences.

Road Building

A good entrance road into your property will be essential from the beginning for delivery of materials and access by workers. Since log homes are often delivered on long-bed tractor trailers, entrance road building may be more complex than for a conventional house. An alternative to building a road to accommodate the log trucks (often 48-foot trailers) is to plan to off-load at some distance from the building site and bring logs in on a shorter truck or by forklift. Whichever method you choose should be reflected in your job-site plan.

Entrance roads must be capable of handling traffic in all kinds of weather, so a simple bulldozer path through the woods will usually be insufficient. The roadbed should be prepared by removing trees, brush, and surface litter. It should be graded smooth, with low areas filled and packed and high points graded down. Then apply gravel. At Laurel Lodge, we laid out the road about four times, trying to find the gentlest slope up the hill, preserve the most trees, and allow wide sweeping curves that would prevent vehicles from meeting blindly. With the roadbed graded, I put down a base of several inches of 2-inch gravel. The loader operator packed this coarse stone with repeated passes before applying a top 4-inch layer of crushed stone ("crusher run" in local quarry language). This, too, was packed to form a hard, all-weather surface. This isn't the only way to make a road. Local road builders can suggest alternatives for different environments.

Well Drilling

As soon as there is good access, I recommend drilling a well. In some situations it may even be possible to drill the well before the entrance road is complete. Well-drilling procedures will vary with different areas of the country; many areas require that the flow rate of the well be tested as soon as it is complete. It's best to allow several days for the entire operation.

Septic System

The septic system may also require clearing, and some sort of roadway will be needed to get the septic tank into place. If possible, try to get the septic system installed while all the other sitework is being done. A septic system generally requires use of a backhoe; it may be possible to use the same contractor you are using for excavation or footers. By completing all of this work at once, you eliminate bringing heavy equipment on and off the site repeatedly. Not only do you save money (in the form of "haul" or "drop" charges), you save the aggravation and delays of having everyone stop work to move their vehicles so the equipment can pass.

PRECONSTRUCTION

Permits

You can generally pick up an information packet on permit requirements from the local building permits office of county or city government. This should explain what permits will be required, how much they will cost, where application should be made, and any inspection procedures that are attached. Permit requirements and procedures vary widely (and wildly) from county to county. If you are planning to build on the East Coast or in California, prepare to learn a whole new meaning for the term "red tape." In less developed areas, permit procedures may be much simpler.

The permit procedure from application to approval may take any-where from hours to months. In most cases, officials will need to review copies of your final blueprints. Be sure to take permit time requirements into consideration when scheduling. There aren't many places left where you can simply go down to the permit office, write a check, and head for your job site to meet the excavator.

Here, as an example, are the permits required in the building of Laurel Ridge and the offices that administered them:

Well drilling	County Health Department
Driveway entrance	State Highway Department
Minor grading	County Sediment Control
Building	County Building Department

Septic	County Health Department
Plumbing	County Building Department
Electrical	County Building Department
Well use	County Health Department
Occupancy	County Building Department

Although many permits are handled by the appropriate subcontractor, the general contractor is responsible for insuring that the proper permits have been issued before any work is done. In many areas, inspectors have the authority to stop work and even require that work be dismantled or redone if proper procedures were not followed. To avoid undue aggravation, contact your permit authorities, visit with inspectors, and make it clear by your enthusiasm to comply that you are interested in the same things they are: quality and safety. You will get more accomplished faster by being on friendly terms with your building permits and inspections office than by trying to circumvent regulations and neglect procedures.

Scheduling

Once your building permit is in hand or imminent, start contacting people on your schedule sheet. Give firm dates that you need certain things done. Whenever possible, avoid tentative scheduling, because it leads to subcontractors considering your project a "filler" or lower priority. Give subcontractors enough lead time so they won't delay progress. It isn't reasonable to call a sub and say, "I got my building permit in the mail today. I need you there first thing tomorrow morning." Most subs require at least several days' notice. When you talk with subs, get a commitment from them for a starting date for their work. Inform them that you consider this a commitment, note it in your construction notebook, and call the subs a few days before their scheduled commitment to remind them. If they don't show up, call and remind them again of their commitment, and ask for an explanation and a final commitment. Tell them if it happens again, you'll have to call your backup subcontractor.

House Layout

Now is the time to lay out your house, if you haven't done so already. In some areas this must be done by a licensed surveyor. If you are building close to property lines, set-backs, or rights of way, you may want to have a surveyor stake the house whether it's required or not.

Material Storage Layout

When you lay out your house, mark out the areas to be used for material storage. This will prevent them from becoming parking lots or debris piles. Make sure that the material can flow smoothly into the house. If possible, lay out material storage with your carpenter and foundation contractor. They can provide valuable insight on how to maintain efficiency.

Excavation

I strongly suggest being present during excavation. Besides the excitement of seeing the first bucketful of earth come from what will be your log home, you will set the tone for the entire project. Workers will expect you to check up on them. If the excavator does not use a builder's level to determine their cut (some use pocket levels for eyeball approximations), you should consider renting one yourself. Use it as described earlier (page 103) to make sure that your excavation goes to the proper depth and that the floor of the excavation is level and smooth. An irregular or out-of-level excavation means that additional fill material or concrete will be necessary to get a level basement floor.

Once excavation is complete, lay out footers. I use spikes or large nails to mark the outside corners of the foundation. Running a string line from corner to corner creates an outline of the foundation. Then I use marking lime (available at masonry suppliers) sprinkled from a coffee can to mark the edge of the footer. Since the foundation wall should sit in the middle of the footer, place the footer line outside of the string line the appropriate distance to center the foundation wall. This distance will depend on the width of the footer and the thickness of the wall. To center the foundation wall, subtract the thickness of the wall from the width of the footer and divide by 2 (so half the footer will be outside the wall and half inside). For example, to center a 10″ block wall on a 24″ wide footer: $24″ - 10″ = 14″ \div 2 = 7″$. So, the edge of the footer should be 7″ outside the foundation wall line. Use lime or chalk dust to mark a solid line around the footers at this distance. The backhoe operator will use this line as a guide when digging his trench. Watch the depth of the trench to make sure it meets building code requirements, but is not too deep. Digging beyond the depth required only adds more concrete expense. Local building authorities and excavators can provide the required width and depth of footers for your area.

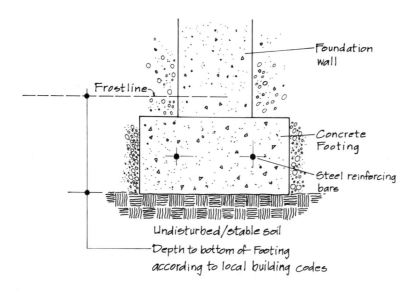

Foundation Wall

Frostline

Concrete Footing

Steel reinforcing bars

Undisturbed/stable soil

Depth to bottom of Footing according to local building codes

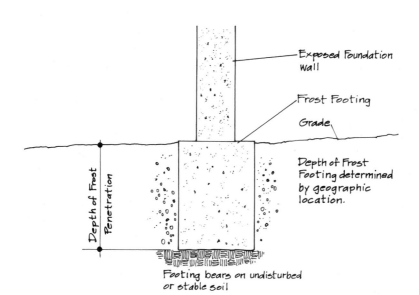

Exposed Foundation Wall

Frost Footing

Grade

Depth of Frost Penetration

Depth of Frost Footing determined by geographic location.

Footing bears on undisturbed or stable soil

FOUNDATION WALL FOOTINGS

In situations where foundation walls will be above ground level, as in the case of a walkout basement, footers must be deep enough to rest on soil that never freezes. This "frost depth" varies with climate conditions throughout the country. In many areas, frost depth exceeds the depth required for conventional footers. In such cases "frost footers" is a term often applied to the extra deep footers required under exposed foundation wall areas. The depth requirements and conditions under which frost footers must be used are available from local building authorities or from excavators. Throughout much of the country, basement garages and walk-outs will require frost footers, and care should be taken that they meet depth requirements. In my area, a building inspector checks the depths of all footers before any concrete can be poured.

FOUNDATION

Once the footers are dug and poured, corner pins can be placed marking the location of the foundation walls. These pins (I use small masonry nails) should be placed accurately using the procedure described on page 99 to check squareness (I don't accept deviations greater than ⅛-inch). Often the masonry or foundation contractor will place the corner pins. If you set the pins, make sure that someone verifies your accuracy.

If you are using masonry block or stone foundation, check the straightness of walls by sighting along their distance. Use a level to check plumb at random intervals and don't be bashful about asking a sub to correct mistakes. On Laurel Lodge, masons had to dismantle and rebuild a 6-foot length of wall in an area of an angled corner. The angle had thrown them off, and the wall started to tilt outward. You can check concrete forms for straightness by sighting along them also, although deviations are harder to see.

Check the top of the completed wall for level, using a builder's level or water level. Any deviation of more than ¼ inch requires leveling with mortar or portland cement. Your goal here is to insure a level subfloor and a draft-free seal between foundation wall and sill plate.

After the foundation walls are up, masonry walls should be parged. Parging is an application of portland cement troweled over the entire outside surface of the block. Because masonry block is very porous, parging is important to getting a waterproof foundation. Also, masonry block walls do not resist lateral forces as well as poured concrete; the parging lends some structural support as well.

Parged masonry walls and poured concrete walls should be waterproofed. A number of waterproofing compounds are available, with black asphalt or tar being a minimum. If you are very concerned about maintaining a dry basement, you may want to use one of the concrete waterproofing compounds and wrap the foundation in heavy plastic. Your subcontractor or building inspector can offer suggestions for your specific area.

In most situations, some provision must be made for carrying water away from the base of the foundation. Water pressure building along a foundation wall during prolonged heavy rains can lead to a flooded basement and can even collapse a wall. In most instances, perforated plastic pipe is placed around the foundation with a drain pipe leading to the ground surface some distance from the house. Some contractors place the pipe inside the foundation wall under the slab; others put it outside around the footer. Discuss this with your contractor and check the opinions of builders in your area. The drain tile should be covered with filter paper to protect perforations from becoming blocked by soil. In many areas, a building inspector must examine drain tile and waterproofing before backfill can take place. In any event, it's good for your peace of mind to see how the drain tile is installed and make sure that it is not damaged or obstructed by debris.

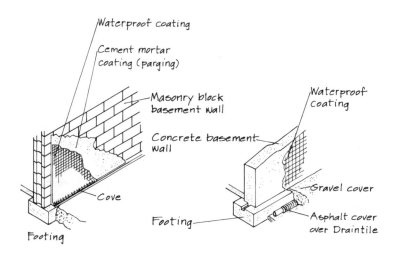

FOUNDATION CROSS SECTIONS

Backfill

Where backfilling falls in your schedule depends on the type of foundation wall and the methods of your subcontractor. A masonry block wall should cure for a while, during which time the subfloor can be added. Many contractors also want the basement floor poured before backfilling. Masonry walls will not handle great lateral pressure, so backfilling must be done carefully. Your contractors' judgment may be helpful here.

Concrete walls can be backfilled earlier, sometimes within a day or two of removing the forms. Often backfilling is done before the subfloor or basement slab is in place. There are many advantages to backfilling soon after the foundation walls are up and before carpentry work starts: you won't have to worry about trash and sawdust accumulating in your backfill trench (where it might damage or clog drain tile); you will eliminate the

PHOTO BY JIM COOPER

Backfilling the foundation should take place as early as practical. I like to backfill before logs arrive because it reduces the risk of injury from someone falling into the excavation, creates more storage area close to the house, and allows more time for settling of the soil before final grading. To avoid the possibility of cracking the foundation wall, no vehicles or material storage should be allowed on the backfill.

potential safety hazard of carrying logs and tools across the open trench; work will proceed faster; and your job site will be neater.

BASEMENT SLAB

The basement slab or floor is not in the critical path of other activities, so pouring is sometimes postponed until after the house is under roof. This schedule assures that the basement area can drain while the upper floors are exposed to weather. Also, the slab will be spared the gouges, smears, and stains that sometimes show up while other construction work proceeds. Alternatively, waiting to pour the slab may mean a mess later on either in the yard or in the house, if a concrete chute must be run over the subfloor or through a first-floor window or door opening. It can mean having to touch up the grading job around the house with additional equipment and operator expense. On a poured-foundation-wall basement you can pour the slab as soon as the walls are finished. On masonry block walls you should wait until the subfloor and some of the exterior walls are erected to help stabilize the basement wall. It's a good idea to brace block walls and inspect for cracked mortar joints immediately after backfilling. On pre-cast foundation walls, it may be a requirement to pour the slab before backfilling, in order to secure the bottom of the wall. In any event, it generally doesn't hurt to pour before backfilling no matter what type of wall is used.

To give the slab some protection from dirt and stains and to give the floor a nicer appearance, I apply a coat or two of an acrylic concrete sealer, available from masonry supply stores. This will give the floor a sheen and prevent stains from soaking into the concrete. I apply the sealer as soon as the floor is walkable, using a long-handled paint roller. It's an easy task, but make sure you have good ventilation, since the sealer contains some potent vapors.

Before pouring the floor, make sure any subsurface work is done. If you plan on having a bath in the basement or simply want to rough-in for future possibilities, plumbing work must be in place before the floor is poured. If you are putting sump crocks, radon loops, or drain tile under the floor, they should be all complete before prep work begins on the slab.

Preparation for the slab consists of leveling the basement floor area with gravel and raking until the base is smooth. Then the area is covered with a layer of thick polyethylene. This is topped by a layer of wire mesh and perhaps rebar, and tied with a twist of wire wherever two pieces cross. If any areas of the floor need reinforcing, such as for a basement garage, you can add a grade beam or two for extra strength. Grade beams are nothing more than thickened areas of concrete reinforced with steel rebar. They can be made by scooping a trench several inches deep and a foot or more wide in the gravel base and laying in pieces of steel. This can give additional strength to areas that must bear excessive weight. If your slab is going to rest on fill dirt, grade beams can be even more important. Fill can settle, leaving a void over which the slab may crack. In many areas, building codes require grade beams over fill to be designed or approved by a licensed structural engineer.

It takes skill and elbow grease to get a good finish on a concrete floor. On areas that can get wet, such as porches, a broom finish (made exactly

PHOTO BY JIM COOPER

Plumbing that will be covered by a basement floor must be installed before the basement floor slab can be poured. This will require coordinating the plumber, mason or concrete contractor, and possibly the carpenter. The foundation walls shown here are pre-cast concrete walls that are set in sections and bolted together. Such walls have footers built-in and rest on a base of fine gravel. Note the pier holes in the center of the foundation area. These will be filled with concrete to support the steel posts that in turn will support the main girder of the house.

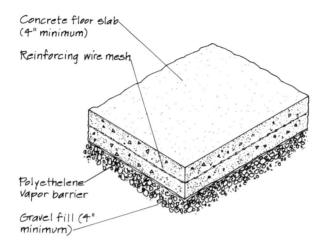

Concrete floor slab
(4" minimum)

Reinforcing wire mesh

Polyethelene
Vapor barrier

Gravel fill (4"
minimum)

BASEMENT SLAB CROSS SECTION

as the name implies, by sweeping the drying concrete with a broom as a final finishing step) can give a non-skid surface. For smooth floors, repeated trowelings as the concrete sets give increasing smoothness.

"Cold joints" result when pouring is interrupted too long between truckloads of concrete. Make sure the concrete contractor orders additional truckloads early enough to prevent previously poured concrete from starting to set. Most concrete subs have a phone or radio-equipped truck, but it's a good idea to confirm that there will be communication available at the job site to contact the concrete supplier without delay.

Concrete may contain additives to help speed or slow drying in extremely hot or cold weather. In colder climates, calcium is often added to speed drying and prevent the possibility of freezing. Ask the sub if he plans to use any additives or if any are necessary. Usually, they don't need to be reminded, but you don't want to risk living with the results of someone else's absentmindedness.

FRAMING

For most people, watching the logs go up is probably the most exciting phase of building a log home. Some groundwork for this stage of construction will insure that work proceeds smoothly and without interruption.

Kit Delivery

I try not to have the log kit on site until the foundation work and backfilling are complete. It's time consuming and aggravating trying to work around open trenches and piles of dirt. Also, if logs and other materials are stacked around the site during site work, you may wind up with some nasty dents or broken materials. Use the job-site plan to determine where materials will be stored when they arrive on site. Your carpenter's assistance will be invaluable here. Try to place logs as close to the house as safety and the site will allow. I try to place logs close to a sliding glass door opening to avoid being "walled off" as work progresses. Determine which materials will be used first and put them closest. Items like shingles and windows can be stored further away or out of the main pathway because they won't be needed until later in the process.

Protect building materials with plastic tarping or by placing them in an enclosed trailer. Don't stack interior trim and flooring outside where they can pick up moisture; covering with a tarp is generally insufficient protection for these items. I know of one builder who builds a "dry room" in the basement as soon as the subfloor and slab are complete. He uses heavy plastic rigged like a large tent and raises everything up off the floor. In winter, he uses a construction heater to keep materials dry and thawed. Many glues and sealants won't work properly at extremely low temperatures, so give these special consideration.

Doublecheck all materials on arrival against the shipping list and the materials list for the house. Inspect for shipping damage or defects as materials are unloaded. (I check all deliveries as a matter of company policy and have the home owner sign for materials.)

Unloading a log home generally requires a forklift. Who pays varies from company to company, but you should make sure that the proper equipment is on site when needed, with people who can operate it. Most trucking companies allow a specified amount of time for unloading. Excessive time is billed at a set hourly rate. I always recommend to my customers an extendable boom forklift (forks are on the end of a boom that can be extended and retracted as well as raised and lowered) over a mast forklift (straight up and down only). Whichever type you choose, it should have rough-terrain tires. The small-wheeled forklifts used in warehouses can't handle the bumps and ruts of a job site.

Several times I've had customers ask if they could unload with a front end loader or backhoe equipped with forks on the bucket. I used to caution against it; now I simply say no. These makeshift rigs will cost any possible savings in time and aggravation. In addition, they cannot be controlled precisely, so there is a real risk of damaging materials. The extendable boom forklift will take much of the strain out of unloading. It is easy to operate, can be precisely controlled, allows you to unload a truck from one side (rather than having to access both sides of the truck or turn it around, which can present more problems than is apparent on many log home job sites), and the boom makes it possible to set logs exactly where you want them, even reaching over mud holes and tire ruts. (It always rains or snows on or immediately prior to delivery day, without exception.)

Subfloor

Most log homes are conventional subfloor systems with either dimensional floor joists or floor trusses supporting a covering of tongue-and-groove plywood. Log kits may or may not include the subfloor system; if it is not included, it is advantageous to install the subfloor, using locally purchased materials, before the log kit arrives.

Framing a subfloor is straightforward, and any general home construction book can give details. Generally, the system consists of a sheet-like foam insulation called sill sealer laid atop the foundation wall. Anchor bolts set by the foundation sub are used to fasten a sill plate made from treated lumber. Floor joists rest on the sill plate, and ¾-inch plywood or composition board is nailed and glued to the top of the joists.

My main concerns are that the subfloor be level and square. I check the top of the foundation wall with a water or builder's level to determine whether the subfloor will need to be shimmed. If the surface is uneven or slopes more than ¼ inch, I have the wall leveled using portland cement. (This is something I check, because many carpenters I've encountered seem to prefer setting the subfloor regardless of level and blaming irregularities on the foundation contractor.) Minor irregularities can be corrected by shimming under the sill plate, using steel shims or treated wood.

Use the procedure described under "Framing Square" on page 99 to confirm that the subfloor is square. If the subfloor is out of square, corner joints won't be tight unless individually cut (precut kit users may have difficulty getting corners to fit). It also means that interior partitions will be much more difficult to lay out, and floor and wall coverings may present problems. Check the level on the completed subfloor deck, having the carpenters shim any low spots between subfloor and foundation. It's especially important to pay attention to the height of any support beams. A

COURTESY GASTINEAU LOG HOMES

The first-floor deck or subfloor rests on top of the foundation wall. A "sill plate" of pressure-treated lumber and a thin layer of sill sealer insulation separate the wood of the subfloor from the masonry or concrete.

string line attached to the sill plate and pulled taut between opposing walls will tell whether beam height is accurate. As the carpenters pull floor joists out to lay in position, you may note them sighting along the board and marking one edge or the other. They are looking for "crown," a slight bow that frequently develops in dried dimensional lumber. The crowned or humped edge should be placed up to create a slight arch rather than sag to the subfloor. With time, the subfloor will flatten itself out.

When using floor trusses, make sure there are working drawings showing how the trusses are to be secured. Some types of trusses have a top and bottom as well as a front and back end. The trusses are engineered for specific placement, and failure to install them properly can result in a sagging or noisy floor.

Before the plywood goes down on the subfloor, it's a good idea to review the blueprints. Interior bearing points or walls may require double

An alternative to conventional floor joists is floor trusses. This web-style truss allows a greater span between supports, which may mean fewer support posts in the basement. Truss-supported subfloors can often be assembled faster than subfloors that use dimensional floor joists. Trusses generally have a front, rear, top and bottom so pay attention that manufacturer's instructions are followed.

Most subfloors are sheathed with tongue-and-groove plywood or composition board. Sheathing should be glued as well as nailed or screwed into position to eliminate future squeaks. Before sheathing goes on the subfloor, framing should be squared to insure that it conforms with blueprints.

or even triple joists, and it's not difficult to forget to go back and double up joists wherever it is called for. It's much easier to add joists before the plywood goes down than to have to wedge and pry additional boards into place later.

Logs

Procedures for setting logs should be spelled out clearly in the instructions supplied by the manufacturer. These should deal not only with placement of the logs and corner fitting, but with sealant installation and methods and spacing for fasteners (spikes, throughbolts, lag screws). Check periodically for straightness by sighting along a log wall. Most log wall systems should be braced firmly as the wall goes up. Braces are usually 2 x 4s fastened to the sides of window or door openings and anchored to the subfloor over a joist. Braces are not removed until the second floor or roof system is in place.

Before log setting starts, I call in the electrical contractor. We go over the location of future outlets and switches in the log walls. I use a red lumber crayon to mark locations on the subfloor. That way we can drill wire holes and cut outlet and switch boxes at the right locations. Not all log home companies include an electrical diagram with blueprints, and those that do may not consider idiosyncrasies of local electrical codes. By working with the electrical contractor I am assured that we won't complete log work only to find that outlets will have to be added or moved.

Pay close attention to blueprints during log raising. Sometimes long walls require permanent bracing or "strongbacks" to prevent them from rippling or bowing from top to bottom. Watch squareness in corners especially, and periodically sight down the corner from above to make sure the wall isn't winding or twisting. Keep an eye on carpenters to make sure they are following manufacturer's instructions on number and spacing of fasteners and placement of sealants and caulk. The key to a weathertight house lies with attention to detail during log raising. Failure to seal or fasten properly can result in recurring headaches for which there are only temporary cures.

Many manufacturers include in their log system construction details designed to accommodate settling or shrinking of logs. Pay strict attention to these. Periodically verify the carpenter's measurements for settlement spaces. Make sure that any required shims are placed properly and that adjustable columns or jacks are correctly installed.

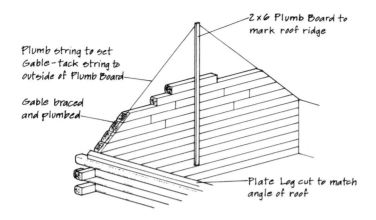

2 x 6 Plumb Board to mark roof ridge

Plumb string to set Gable - tack string to outside of Plumb Board

Gable braced and plumbed

Plate Log cut to match angle of roof

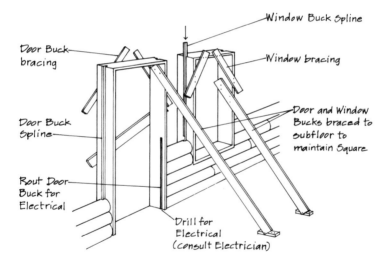

Window Buck Spline

Door Buck bracing

Window bracing

Door Buck Spline

Door and Window Bucks braced to subfloor to maintain Square

Rout Door Buck for Electrical

Drill for Electrical (consult Electrician)

BRACING DETAILS

(COURTESY GARLAND HOMES)

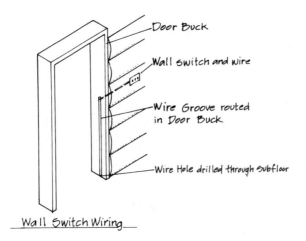

Door Buck

Wall switch and wire

Wire Groove routed
in Door Buck

Wire Hole drilled through Subfloor

Wall Switch Wiring

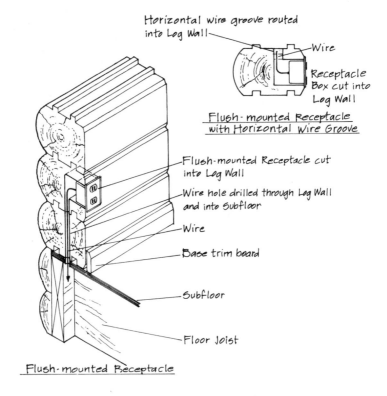

Horizontal wire groove routed
into Log Wall

Wire

Receptacle
Box cut into
Log Wall

Flush-mounted Receptacle
with Horizontal Wire Groove

Flush-mounted Receptacle cut
into Log Wall

Wire hole drilled through Log Wall
and into Subfloor

Wire

Base trim board

Subfloor

Floor Joist

Flush-mounted Receptacle

RECEPTACLE WIRING

COURTESY GASTINEAU LOG HOMES

Methods of laying log walls vary among companies. The manufacturer's construction manual should spell out procedures clearly. Regardless of the method used, a neat worksite and efficient use of manpower are essential for both safety and quality. Note the bracing supporting the door buck. Company bracing instructions should be followed precisely to avoid crooked or bowed walls.

Exterior Framing/Sheathing

On many log homes, gables, dormers, and bay windows may be framed conventionally rather than using solid logs. Exterior framing is usually a 2 x 6 covered with sheets of insulated sheathing or plywood. When using insulated sheeting it's still a good idea to sheath corners with plywood for extra strength. Siding generally goes on after the roof system is complete.

Roof System

Roof systems differ markedly among log homes. I group them into two categories: (1) conventional framing uses dimensional rafters or trusses topped by plywood, tar paper, and shingles; (2) timber framing uses heavy

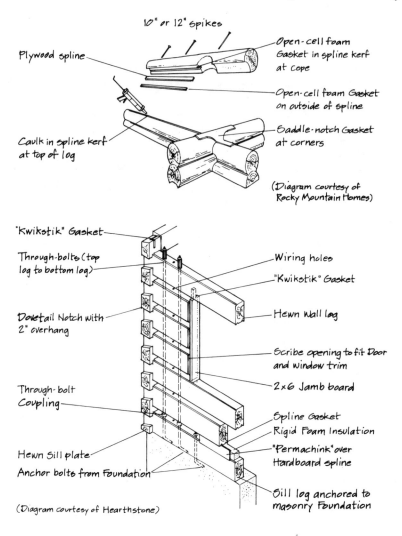

10" or 12" Spikes

Plywood spline

Open-cell foam Gasket in spline kerf at cope

Open-cell foam Gasket on outside of spline

Caulk in spline kerf at top of log

Saddle-notch Gasket at corners

(Diagram courtesy of Rocky Mountain Homes)

"Kwikstik" Gasket

Through-bolts (top log to bottom log)

Dovetail Notch with 2" overhang

Through-bolt Coupling

Hewn Sill plate

Anchor bolts from Foundation

(Diagram courtesy of Hearthstone)

Wiring holes

"Kwikstik" Gasket

Hewn Wall log

Scribe opening to fit Door and window trim

2x6 Jamb board

Spline Gasket

Rigid Foam Insulation

"Permachink" over Hardboard spline

Sill log anchored to masonry Foundation

WALL CONSTRUCTION

timbers for the structure of the roof, topped by wooden decking (usually 2 x 6 tongue-and-groove), insulation (often sheets of styrofoam or polyiso-cyante), a nailbase sheathing, tar paper, and shingles. Both systems have their advantages, although most people prefer the appearance of a timber roof. (A conventional roof may include heavy timbers, but they will be decorative rather than structural.)

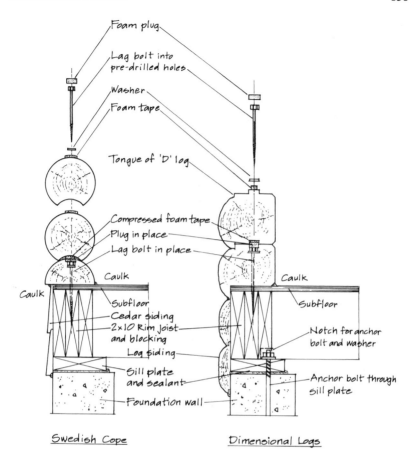

Foam plug

Lag bolt into
pre-drilled holes

Washer

Foam tape

Tongue of 'D' log

Compressed foam tape
Plug in place
Lag bolt in place

Caulk

Caulk

Caulk

Subfloor
Cedar siding
2x10 Rim Joist
and blocking
Log siding

Subfloor

Notch for anchor
bolt and washer

Sill plate
and sealant

Foundation wall

Anchor bolt through
sill plate

Swedish Cope

Dimensional Logs

FASTENING LOGS

(COURTESY GARLAND HOMES)

Follow manufacturer's instructions closely and determine beforehand any special equipment, such as cranes, that will be needed. Carpenters experienced with heavy timbers can work wonders without resorting to cranes and expensive hoists, but don't wait until roofing day to find out what's needed.

Things to watch for include spacing and placement of rafters. Make sure cuts are neat and accurate for valley and hip rafters. For some carpen-

When log walls reach second-story height, it's time to set the girder and beams to support the second story. The post supporting the girder in this photo actually does not rest on the subfloor but passes through it to rest on an adjustable jack that sits atop the main girder beneath the subfloor. This allows adjustment of the first floor girder in response to any log settling. Note the notch cut into the girder for second floor beams.

A portable hoist speeds the setting of second-story beams. Rental of such special-ized equipment should be included in either the carpentry subcontract or in a miscellaneous budget.

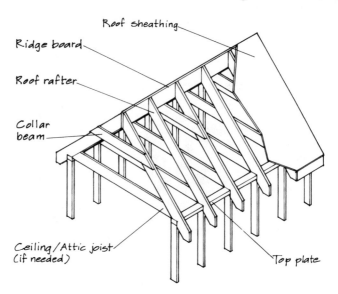

Roof sheathing

Ridge board

Roof rafter

Collar beam

Ceiling/Attic joist (if needed)

Top plate

Conventional Roof Framing

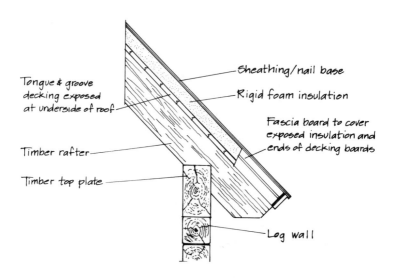

Tongue & groove decking exposed at underside of roof

Sheathing/nail base

Rigid foam insulation

Fascia board to cover exposed insulation and ends of decking boards

Timber rafter

Timber top plate

Log wall

Timber Framing

ROOF FRAMING

ters, cutting compound angles borders on magic, and it is sheer luck when such cuts fit tightly. A carpentry crew that I no longer use apparently had one carpenter who had mastered the intricacies of compound miters, while the others had not. The house we were building had several paired valleys. On each, one in the pair fit nicely, while on the other an angled cut hadn't even been attempted. Instead, they just butted the two intersecting members and put several pounds of nails into the ugly results. I spent an afternoon nailing metal plates to strengthen those joints (the offending crew was unemployed by then).

Log homes often have large ridge beams. A crane is used to set this 44-foot laminated beam, with ropes guiding it into position. Note the vertical brace supporting the log gable. Such bracing is extremely important because log gables are supported only at their base. A strong wind can topple an unbraced gable.

Methods of roof framing vary by manufacturer and with homeowners' budgets and desires. Here a conventional roof system is used. The dimensional rafters will be covered with plywood, tarpaper and shingles. The spaces between rafters will be filled with fiberglass insulation. Note the framing that supports the roof over-hang. Notching the log gable and extending cross braces or "lookouts" from the flyrafter on the outside to a rafter inside eliminates the risk of sagging eaves.

COURTESY GASTINEAU LOG HOMES

Truss roofs are becoming more common. They are faster to construct and the engineered trusses allow greater spans than conventional rafter roofs. This scissor truss is designed to create a cathedral ceiling in the greatroom it covers. The underside of the truss will be covered with tongue-and-groove pine to create an attractive wooden ceiling. A truss roof such as this is usually available either as an option from the log manufacturer or from a local truss supplier or lumberyard.

Interior Framing

Usually, only load-bearing partitions are installed when the exterior of the house is complete. This leaves the interior relatively open, allowing freedom of movement with long and awkward materials. After the roof is finished, interior partitions can be installed. Check to make sure rooms are square and partitions plumb. Check squareness and dimensions of door openings. Make sure provisions for settling are properly installed.

Now is the time to frame recessed medicine cabinets. You'll kick yourself (unless your carpenter does it for you) if you close up a wall that must be ripped open and butchered to allow a recessed medicine cabinet. You'll kick yourself even harder if you rip open a wall to find wiring or plumbing in your way. Check dimensions or bring the medicine cabinets to the job site to block out the recess accurately.

Framing partitions can take place before or after the roof is set. Since a truss roof may eliminate the need for a center girder, most partition framing can take place after the house is covered. In this case, however, good weather made it advantageous to frame the interior first.

Another builder friend once suggested inletting or notching a horizontal 2 x 4 called a "nailer" into walls that will receive hanging cabinets. Locate the nailer at the proper height to receive the cabinet screws. Your cabinet installer won't have to measure and poke around (risking false holes) to be sure there is a stud behind the cabinet when fastening them in place. It's a simple trick but contributes to better workmanship all around.

Windows and Doors

Window and door installation is fairly simple using modern preassembled units. Building doors from scratch requires skills beyond many modern carpenters. If you will be building windows or doors from scratch, make sure they are within your carpenter's capability before you turn him loose on good lumber.

Most window units are simply set in place in a "window buck" (a rough frame of dimensional lumber attached to the logs. Some manufacturers request or require that bucks be attached to logs with nails or screws passing through slots rather than holes. This allows logs to move without altering the position of the window. If your log kit requires such methods, be sure the carpenters actually use them. Failure to do so may prevent logs from settling, causing cracked frames or windows and doors that don't open and close tightly. Nail slots are easily made with a router, or by plunging a circular saw through the lumber with the blade parallel to the board.

When framing door openings, your carpenter will need to know what type of floor will be under the doorway, so he can make the opening the proper height. You should have your flooring scheme in mind before he starts framing.

Exterior Trim

Exterior trim is not only important to the appearance of the house, it contributes to weathertightness and longevity. Shoddy exterior trimwork can lead to air or water infiltration around doors or windows. This will wreak havoc with utility costs, and it may result in areas of rot or mildew. Make sure that all exterior trim is neat, with tight joints and that caulk, foam sealants, and flashing are used where specified by the log kit manufacturer. Metal flashing is often used above windows to prevent water from seeping between the window trim and the log wall.

ROOFING

Shingles

A variety of materials can be used to cover the roof of your log home. Slate, cedar shakes, tile, and metal all can be used just as with a conventional home. However, standard fiberglas or composition shingles are supplied most often with log kits.

When using standard tab shingles, watch the work to make sure that it is laid out properly. Tabs should be spaced regularly to give the roof a neat, symmetrical appearance. Improper or uneven spacing of tab shingles is noticeable immediately.

Check the method used to finish valleys in the roof. Poorly finished valleys can lead to leaks. Valleys occur where sections of pitched roof with separate ridges intersect. The valley forms a channel that collects water.

The easiest method of finishing a valley is to install a continuous run of metal flashing. Shingles overlap the edges of the flashing but do not conceal it. Shingles may also be overlapped or woven over valleys. These add further protection when the valley is carrying water.

Vents and Drip Edge

Roofs can collect a tremendous amount of heat. To eliminate excessive heat buildup, conventionally framed roofs use a vent system. Eave or soffit vents (strips of metal or wire mesh) allow air to enter the roof under the eaves. The air rises along the underside of the plywood roof sheathing to a ridge vent, where it is released. Heavy timber roofs that use solid styrofoam or stress skin panels will not have a vent system, since there are no air spaces. Ridge vents can be exposed metal venting that forms a small cap on the ridge, or they can be concealed strips that lie under the ridge shingles raising them enough to allow air to circulate.

At the edge of the roof a metal strip that covers the edge of the plywood sheathing is installed under the shingles. This drip edge provides support, allowing shingles to overhang the roof edge slightly, and prevents water from soaking the edge of the plywood roof sheathing, a condition that will lead to decay.

Soffits

The underside of roof overhangs and eaves is called the soffit. Different log home manufacturers have different soffit finish materials and methods. Some companies provide no soffit covering. Soffit materials are usually thin plywood or strips of tongue-and-groove pine or cedar. The soffit covering of a conventional roof generally contains a vent (described above) to allow air circulation beneath the roof.

Study the soffit closely, making sure that it covers roof overhangs completely and fits snugly against the outer edge of the roof (fascia or subfascia) and against the house. This may be difficult where soffits cross rounded logs at an angle, leaving spaces where the logs narrow. Openings such as these should be caulked, trimmed, or otherwise sealed to keep out insects. As soon as most log homes are complete, the word goes out among flying critters to come and have a look. You will notice the house being inspected by bees, wasps, flies, moths, and birds. If they find an entry into the roof system, you could be in for problems. Some exterior sealants contain insect repellents that discourage them from visiting such areas.

Gutters and Downspouts

These serve a purpose beyond keeping water from running down your neck when you come home during a heavy rain. The gutter system functions to collect water and remove it from the immediate vicinity of the foundation. Gutters should slope toward downspouts that carry water down and away from the house.

Splash blocks should be used to prevent water from hitting the ground directly as it exits the downspouts. This prevents gullying and the formation of holes that can carry water down around the foundation. In some instances, I run downspouts into underground plastic drain tile (unperforated until it is well away from the house).

When a log house has wide roof overhangs, an owner may choose to forgo gutters and downspouts, instead placing a gravel-filled trench around the house to absorb water. The trench is separated from the house and should contain underground drain lines leading away from the foundation.

The ultimate goal of any roof drainage system is to collect water and get it away from the house. Prolonged dampness is the number one enemy of a log home, virtually guaranteeing decay. It also can cause damp or wet basements and, in extreme cases, lead to failure of the foundation walls.

FINISHING

Weatherproofing and Caulking

The energy efficiency and structural soundness of your log system is greatly affected by the sealant system used by your manufacturer and how it is installed. The sealant system usually consists of a combination of foam or butyl gasketing, expanding foam, splines, and caulk. Read the manufacturer's instructions for using these seals and make sure that your subcontractors follow instructions to the letter. Be open to suggestions or warnings from subs who may notice a potential problem that isn't covered in the manufacturer's standard construction details. Study your log home constantly as it's being constructed, asking yourself if the finished construction will handle water and air infiltration.

Insulation

This is usually installed by an insulation subcontractor. The subcontract should specify that the work will conform to local building codes. Some codes specify the amount and type of insulation that can be used in certain situations. For example, standard fiberglas batts faced with kraft paper are often used to insulate conventional rafter roofs and attic spaces. The insulation is then concealed by the interior ceiling covering. If a partition is to be insulated but remain unfinished, many fire codes specify that the kraft paper facing must not be exposed. Options include turning the insulation backwards so the paper facing is against the finished side of the partition, or using unfaced fiberglas batts. Either option is superior to racing frantically to rip the paper facing off of exposed insulation before the building inspector fails the house.

Many insulation subcontractors include sealing around door and window framing with expanding foam sealants. If this is included in your agreement, check to make sure that the job is done thoroughly and be sure the contractor knows beforehand what areas should be left unfoamed. (I once had an industrious insulator who foamed several electrical wire holes before I had a chance to warn him. The electrician was not pleased.)

Basement areas get special consideration in building codes. Unfinished basement areas can usually remain uninsulated with insulation being placed between the floor joists beneath finished living areas. In some areas, however, placing heating ducts in unfinished basement areas turns them into finished space as far as code regulations are concerned. In such situations, a blanket of insulation may be required around the basement walls in areas where the basement wall is above grade. This insulation must be fire retardant and so is somewhat more expensive than fiberglas batts. It is attached by nailing into the masonry or concrete.

In many areas, insulation must be inspected before it can be covered up. Be clear about who schedules the insulation inspection.

HVAC

There are many ways to heat and cool log homes. Heating methods vary from woodstoves to heat pumps and furnaces to various radiant heat systems. Virtually any system used in a conventional home is acceptable for log houses.

If your HVAC system involves ductwork, it's best to get the subcontractor involved early in the construction process. Many HVAC subs are used to picking out a handy closet or corner and framing a passageway or "chaise" to conceal ductwork running from the first to the second floor. If your house has open-beamed ceilings and few interior walls, choices for running concealed ductwork between floors will be limited. Let the subcontractor know what to expect before he shows up to run ductwork. It can save you money and the aggravation of unsightly duct placement.

Plumbing

A major difference between modern and historic log homes is the location of plumbing. Today it's inside the home instead of a few yards behind it. Kitchens have moved indoors, too. For the plumber there are few differences between log homes and conventional frame dwellings (unless interior partitions are solid log, also).

Water lines are run through conventional frame walls to service bathrooms and kitchen. "Plumbing walls" are framed walls that carry plumbing such as waste and vent lines between floors and to the roof. A plumbing wall is generally 2 x 6 instead of 2 x 4 to carry the larger pipes. Placement of plumbing in two-story houses is much easier (and cheaper) when bathrooms are located one above the other (stacked).

The main difficulty for the plumber comes in dealing with bathrooms that are above open beamed ceilings. The problem is how to conceal the floor drains for showers or tubs and toilets. In many log homes bathroom ceilings are "dropped" with drywall or tongue-and-groove over conventional joist framing. In other instances, a second-floor bath may have the subfloor built up to create a space for pipework. Remember that the space required beneath a bathroom floor must be greater than just for the pipes alone. Waste lines must be sloped to allow gravity flow, so the space must accommodate 2- or 3-inch pipes that angle at a rate of ¼ inch per foot or more.

Go over your floor plans with the plumber well before he arrives for rough-in. Plumbers are usually scheduled to follow HVAC, overlapping by a day if possible.

Electrical

There are a number of electricians who won't undertake wiring a log home. For some, it's simply fear of the unfamiliar; for others, it's previous experience with a log home in which preparation for wiring was inadequate. Wiring a log home can become a costly and frustrating undertaking if the house has not been properly prepared. Wiring of framed interior partitions and conventional rafter or truss roofs is no different from conventional housing. The difference again is dealing with log walls, open-beamed ceilings, and built-up roofs.

The simplest (and generally least attractive) method for wiring a solid log wall or exposed beam ceiling is wire mold. These painted metal channels cover wires running along the surface. Other than the advantage of making it easy to change or add wiring at will, wire mold has little to recommend it. In certain situations, code requirements may influence the use of wire mold.

A step above wire mold is wiring concealed behind a large baseboard, held away from the wall by furring strips and usually capped with decorative molding. This is the method often used in modernizing historic log or timberframe homes. Outlets are set in the baseboard. Unfortunately, there are still open-beamed ceilings to contend with, and baseboard wire chases offer no help here.

The final (and what most people consider most acceptable) alternative is to conceal wiring within logs and beams. Some companies offer their packages with wire holes predrilled but, because of variations in building codes and individual tastes, most leave it to the owner.

To wire a log home, I start during the blueprint stage of the house. The final set of blueprints prepared by my manufacturer includes a suggested wiring diagram that meets the National Electric Code. On the preliminary blueprints, the home owner and I note the location of outlets and fixtures. By noting preferences on the preliminary drawings, we can incorporate features beyond code requirements (extra outlets, switches, fixtures) into the final plans. This also helps the electrician estimate the cost of the job, eliminating a possibly nasty surprise later on.

As soon as the subfloor is complete, I meet with the electrician to mark the location of holes needed to run wires for outlets and switches. These must be marked with a permanent marker directly on the subfloor, unless your log home company predrilled or routed wire channels in the logs at the mill. This would eliminate the need for drilling and routing on site (although the electrician should still be consulted to make sure that outlet and switch locations meet local electrical code requirements). Otherwise, as the log walls are erected, carpenters drill 1-inch holes at marked locations. Throughout log laying, we keep a long length of stiff wire to run through wire holes to insure that the passageway is unobstructed. Another approach, if the electrician and carpenter are agreeable and the carpenter has experience, is to have the carpenters lay wire in the logs as they are being erected. Then wire holes won't be forgotten or misaligned.

When the walls reach outlet height, carpenters cut box holes into the face of the logs. Sizing of these holes is critical; they must be deep enough to accept the box fully with any wire clamps firmly screwed down (box depth plus about ½ inch), and an outlet plate must completely conceal the outlet box and hole. We keep a trial box on site to check each hole. A two-gang and three-gang box are also on site to lay out and check multiple-switch box holes. (I once visited a log home where the electrical box holes were all cut about a half inch larger than could be concealed by the cover plate. Whoever was responsible should have been shot at the very least.)

When second-floor beams are in place and before tongue-and-groove flooring is installed, I call in the electrician again. Together, we locate fixtures and wire paths in the beams. My carpenters then chainsaw a channel in the top of the beam, and the electrician lays in the wire. A piece of wire mold is placed over the wire to protect against nail punctures when the floor is installed. A 1-inch hole is drilled in the fixture location and the wire end poked through. The other end of the wire emerges somewhere that the electrician can get at it to complete rough-in work.

Note that this system allows wiring the walls after the house is dried-in and other partitions are in place. Wires can be fished from the basement to outlets and switches. In houses built on concrete slabs, this system naturally won't work. In such cases, log walls must be erected to the height of the outlets and a wire channel cut in the top of the log. Wire must be laid in place before log laying can be continued. Once in place there is little possibility for repair or change.

The key to getting your house wired correctly with no unsightly compromises lies in careful planning, good communication with both your electrician and your carpenter, and strict supervision and vigilance on your part. On one house, I felt a sinking feeling as I poked a test wire into a wire hole and felt a solid obstruction about 6 feet down. I kicked myself for not checking more often, while the carpenter sputtered that he was certain the hole was good. On that February day, while the carpenter and I discussed plan B and the need to be more attentive, the sun began to warm the logs. When we went back for a final check, the wire slipped easily through the entire hole, leaving a little plug of ice where it emerged. Feeling greatly relieved, I swore that checking wire holes would become a priority in the future. Now, as each log is added, the log setters poke a length of wire through to assure that there will be no surprises when rough-in wiring time arrives.

Drywall and Interior Wall Coverings
Supervising drywall subcontractors or trim carpenters installing tongue-and-groove is mostly a matter of watching the quality of work. Before beginning work, the house should be thoroughly cleaned and all trash and

debris removed to allow easy movement of people and materials. Floors should be swept clean, with corners and intersections between partitions and floors cleared of obstructions and dirt.

Before beginning wall coverings, walk through the house with the carpenter and make sure there is something to fasten drywall or wood to in each corner. In framing there sometimes isn't a stud or rafter located to catch the corner of a drywall sheet or the ends of tongue-and-groove. In such cases a "nailer" must be added. Good drywall hangers will add nailers as necessary (at additional cost); poor drywall hangers will do whatever they can get away with.

As work progresses, keep an eye on spacing of nails or screws in drywall. Too few fasteners can result in sagging ceilings or rippled walls. Look for tight-fitting joints and neat work that will make finishing easier. On wood walls or ceilings, look for a tight fit and make sure carpenters countersink nail heads. Be clear with the carpenter, beforehand, whether countersunk nail heads should be covered with wood putty.

Finishing drywall joints is an art. Poor finishing results in visible seams and bumps marking nail locations. A strong light shined against the work at a sharp angle will highlight any unevenness. I specify three finish coats on drywall joints. The first or tape coat beds the joint tape, the second or block coat covers the tape, and the third or finish coat gives a smooth surface that will be indistinguishable from the wall.

Finally, specify beforehand who is to clean up. Drywall work generates a lot of scrap. Unless you plan to do the cleanup, make sure that you are satisfied before making final payment.

Paint, Stain, and Varnish

The quality of paint and varnish work will say much about the quality of your log home. It is difficult to convince anyone that quality construction lies beneath a shoddy paint or varnish job. Make sure that quality paint and varnish are used, not materials left over from other jobs. Your subcontract with the painter should specify that all materials are to be left with you at the completion of the job.

Paint and stain work is very much a matter of personal taste (and budget). Painting is generally significantly less expensive than staining and varnishing. It's best to have a "finish schedule" that specifies which areas of the home will be painted and which varnished.

First choice for paint for walls and ceiling is a flat latex paint applied with a roller. A primer coat should be applied first. This may be applied with a sprayer, depending on the amount and location of exposed wood. Kitchen and bath areas are often coated with a semigloss or flat latex enamel. The enamel provides a more durable finish in areas that are subject to moisture. Also, it's easier to remove stains and grease spots from an enameled surface. Wiping with a damp cloth or soapy rag will remove most mars or spots.

There are many natural wood finishes for covering logs, and each has its fan club. The most durable is probably polyurethane (sometimes called polyurethane varnish), which may be clear or tinted. This seals and protects wood, leaving a satin or gloss furniture-type finish. A polyurethane finish should be a permanent finish; it is washable and, unlike paint, should need no recoating. Several precautions must be taken when applying poly. Make sure the house is warm enough. Polyurethane applied to a cold log wall will dry slowly (or not at all) and may run badly. Once dry, the only way to fix runs or drips is to scrape, sand, and recoat the area with a brush. Surfaces to be coated should be completely dry. Dampness in the wood will lead to poor drying and may fog the finish, leaving an overall milky appearance where crisp grain should show.

Some people prefer an oil finish over polyurethane. Oil penetrates the wood rather than sealing the surface. Oil finishes are often used where a rustic effect is desired. The oil accentuates the grain without leaving a surface coat on the wood. Repeated application can bring a deep richness to the wood grain without a surface buildup.

Recently, water-based polyurethanes have joined the ranks of wood finishes. These products are advertised as "environmentally sound" because they don't contain volatile solvents or depend on petroleum products for thinning and cleaning. They are slightly more expensive than traditional solvent-based systems, but the advantages of no fumes, water cleanup, and less toxicity make them worth considering.

Polyurethane and oil finishes are sometimes applied over stain. Before staining, finish a scrap piece of log without a stain to see the effect. Then use scraps to test various stains and be sure to finish over the stain before making a final color choice. Stains may be brushed, sprayed, or wiped on. Usually the surface is wiped with a dry, lint-free rag after application to avoid runs or excessively stained areas. The idea behind stain is to accentuate the natural grain of the wood.

True varnish finishes are sometimes used on logs and wood trim. These oil-base products provide a durable finish but may be more difficult to apply. Care must be taken to provide adequate ventilation to eliminate toxic fumes. Tools must be cleaned with turpentine or mineral spirits. Varnish products may need to be thinned to produce good coverage, depending on the characteristics of the wood and the environment.

Lacquer finishes are common on wood furniture but are rarely used on log walls. Lacquer is usually sprayed on, building a surface with repeated coats. Some lacquers have extremely volatile solvents; any open flame or spark in a fume-filled area may ignite a flash fire or create an explosion.

Inspect the work after primer is applied. This is the time to note the quality of any drywall finish. Bring back the drywall sub to take care of any flaws before the final coat of paint is applied.

As the painters work, keep an eye on the quality of their work. Make sure prime coats are dry before top coats are applied. Make sure excess stain is removed, and that varnishes are not allowed to run. If necessary, rent or borrow a propane or kerosene heater to "pre-heat" the house for several days before paint and varnish are applied (turn off the flame before work begins if flammable vapors are involved).

Trim is usually painted after it is installed. Watch the painters to make sure they mask areas that should not be painted. Be clear about who cleans glass surfaces after painting. Painters often will not mask glass, choosing to scrape after the paint has dried. Sometimes short-term memory loss, possibly caused by breathing paint fumes, causes the painter to overlook scraping windows before he submits his final bill. Be sure to walk through the entire house with the painter before signing a check. (I walk through the house alone or with someone who hasn't been involved in the painting before walking through with the painter.)

Floor Coverings

Log homes use a number of kinds of floor coverings. It's generally not difficult to recognize what to look for in workmanship; the appearance of the work will indicate problems. However, there are peculiarities to watch in the preparation of the floor and the installation of each type of flooring.

Sheet vinyl, resilient and ceramic tile, slate, and stone generally require an underlayment. I like to use ¼-inch plywood under thin materials and ½-inch or ¾-inch for heavier floors such as ceramic tile. The subfloor should be thoroughly cleaned and any major irregularities repaired before putting down underlayment. Underlayment should be nailed (or fastened with drywall screws) and glued using a very tight pattern to avoid ripples in the subfloor. (I use nails ever 6 inches.) The surface of the underlayment should be smooth with no protruding joints. Knots or punctures in the underlayment should be filled with a leveling compound. Check the alignment of seams in sheet flooring and the straightness of grout lines in ceramic tile. Don't be afraid to tell the subcontractor to redo work that is not to your satisfaction.

Carpeting is straightforward. Make sure that the subfloor is swept and the corners thoroughly cleaned and vacuumed. Check seams and look for ripples that indicate insufficient stretching.

Hardwood is generally applied over the subfloor on a base of one or two layers of 15# felt paper (tarpaper). Watch the tightness of joints. Finishing hardwood floors requires special equipment and skills. The floor must first be sanded with successive grits of sandpaper, using a large floor sander and a special edging sander that sands flush to corners. Care must be taken not to produce a wavy appearance caused by uneven sanding. Pine floors are actually more difficult and may be more expensive to get quality work. The softer pine clogs sandpaper faster than hardwood, resulting in more sandpaper being needed and more time spent changing paper. The softness of the wood also means the sanding machine will cut more deeply, requiring more time and skill to produce a perfectly smooth surface.

It helps to have an idea before floor coverings are installed of how intersections of different coverings and doorways will be treated. Some areas may require thresholds, such as between baths and bedrooms. Should the threshold be wood to go with a hardwood-floor bedroom or should it be marble to go with an adjacent ceramic-floor bath? These kinds of considerations should be discussed with your subcontractor preferably before work begins.

Cabinets and Vanities

These finish details require few special considerations in a log home. As in any other kind of home, check to make sure countertops are level and that doors and drawers work properly. Check the alignment on matching doors; sometimes one door will be slightly higher than the other, requiring a minor adjustment in hinges. Also check the installation and alignment of hardware.

Cabinets mounted against log walls require special consideration. They should be able to accommodate settlement in the walls, depending on the log system used. Cabinets may be fastened to the wall through

COURTESY GASTINEAU LOG HOMES

When the house is "closed in," interior finish work can begin. Custom stairs, railings, door and window trim, and base and crown molding complete the carpentry for the house. Here a carpenter sands what will become loft railings.

screw slots or firmly fastened to a wooden spacer or furring strip that is attached to the wall with slotted screws.

Make sure countertop intersections with walls are finished to prevent water from draining between them. Usually this means a backsplash or caulk joint. Caulking should be neat and fairly unnoticeable.

Fixtures

Appearance and feel will tell you what you need to know here. Are towel bars and toilet-paper hangers level and fastened tightly to walls or cabinets? Are light fixtures fastened securely and trimmed properly? Are plumbing fixtures fastened securely and grouted or caulked?

Interior Trim and Stairs

There is a great variability in the kinds and styles of trim for log homes. To avoid grief, make sure that you and your trim carpenter have the same things in mind before trimwork starts. I've been in a number of log homes in which interior trim consisted of plain 1 x 4 or 1 x 6 cedar or pine. Some log home manufacturers supply preformed moldings with their homes. In other situations, molding may be made on site.

Interior trim usually consists of window casings and sills, door casings, thresholds, and baseboards. Closet shelves and rods are part of trimwork, too, as is any special crown mold around ceilings. In some styles of construction, crown mold conceals settlement spaces at the tops of walls. Make sure that your trim carpenter understands what kind of trim will be used and where.

Stairs may be of the premanufactured variety that the trim carpenters simply install or they may be a site-made staircase. If your house requires the latter, check the stair craftsmanship of your trim carpenter before work begins. I once used an excellent trim carpenter who couldn't make stairs worth looking at. I ended up paying to remove a set of stairs and replacing them with a premade set. The skill level required (and the cost) of site-made stairs will depend to a large degree on your taste. A rustic set of stairs — planks or half logs bolted to rough-sawn stringers — is much easier and cheaper than a carefully fitted set of hardwood steps. Go over stair plans in detail before beginning work.

10

Maintenance

AS MENTIONED IN Chapter 1, modern log homes are not maintenance free. Certain types of maintenance are essential to retaining the comfort, safety, and value of your log home. If a log home sales representative, friend, or self-styled expert tells you that the greatest benefit of log home living is that it is maintenance free, a proper response is, "Hah!"

What is different (and favorable from many points of view) about log home maintenance is that is is often simpler, less costly, and less time-consuming than maintenance of a conventional home. Many of the maintenance areas are completed during the first part of the building process. If they are done properly, you are well on your way to a low-maintenance (as opposed to no-maintenance) log home. Some maintenance activities are specific to a particular type of log home, so pay close attention to directions and suggestions included in your construction manual.

LOG CARE INSIDE AND OUT
Proper care of the logs starts the moment they arrive on your job site. Logs should be unloaded in a dry area and kept above ground. Rough lumber or scrap wood can be used to keep the logs away from ground moisture. Plastic tarping is good if you are building in a wet area or season. If you cover your logs with plastic, do it right. Simply throwing a tarp over the

logs and tossing sticks and rocks on top of it to keep the tarp from blowing away is a waste of time. If you get the kind of weather you are trying to protect the logs from, the tarping will blow off, shred, or fill up with water. Instead, stack the logs to permit the tarp to peak, allowing water to drain off. Keep the tarping taut and secure it on the sides and around the base.

I use pieces of wood lathe (narrow wood strips available at most lumber yards) nailed into the log with small duplex nails—these have two heads, so they can be pulled easily and repeatedly. By nailing the wood lathe over the top of the plastic, you reduce the amount of wind that can get inside. You also reduce flapping that can shred the tarp.

Don't leave tarping on during good weather. Plastic tarps hold moisture and can produce slippery, soaked logs; mold and mildew may even appear. Uncover the logs whenever the weather permits, allowing air to circulate freely. Cover logs only at the end of the day or when bad weather threatens.

During the course of setting the walls, logs are often exposed to foul weather as well as the bleaching action of the sun. By the time log work is complete, logs may be dirty, sawdust and mud spattered, and sun bleached. As soon as the exterior of the house is complete, I wash the logs down with a bleach solution. Plain laundry bleach diluted to ½ or ⅓ strength works fine. I spray it on with a garden sprayer, let is sit for a half-hour to an hour and rinse it off. If the logs are heavily soiled or discolored, I may repeat the procedure or use a stronger solution.

Often you will be able to see the bleaching action in progress. Make sure to cover all log surfaces to avoid streaking that results from uneven bleaching. Let the logs dry thoroughly after bleaching. Actually, I've found that the best way to rinse the logs after bleaching is with a high-pressure washer. These are available for rent and will leave the logs looking like they just left the mill.

After bleaching, I apply two coats (second coat applied before the first dries) of a water repellent that contains a pigment and ultraviolet inhibitor. Not only does this protect the logs from further water marking, but it also gives the logs a rich color and prevents sun bleaching. A number of exterior wood treatments are available, and great and sometimes heated debates occur over the advantages of one or another. If one is not included in your log package, consider the recommendations of your log home company or study articles in magazines and books.

Inside-log treatments are more numerous than exterior treatments. Much depends on the look that you are after. Some log home owners do nothing at all. More put on some kind of finish, ranging from oils to stains and varnish. My standard interior finish procedure (which some say is excessive, but I like the results) starts with bleaching the interior log surface (it gets dirty before the roof goes on). When the logs have dried, I sand flat log walls with a power sander (I use a disc grinder with sanding disc attachment). Choice of grit is critical to avoid swirl marks. A medium grit is best with heavy hardwoods, while a relatively fine grit works best with softwoods. On rounded log walls. I quickly hand sand to knock off surface "whiskers."

With sanding complete and the inside of the house thoroughly cleaned of dust and debris, I apply a coat of sanding sealer. (This is preceded by stain, if desired.) This is how the logs will remain until the interior wall coverings are in and any interior partition painting is done. Then I use an electric palm sander with a fine grit paper to smooth the sealed logs lightly. (This process goes very fast. I work the palm sander at about the speed I would wash the walls with a sponge.) Finally, I apply one or two coats of polyurethane varnish. The result is a finish that will last for years. Also, by thoroughly sealing the log surface, I reduce checking as the logs age.

With such an initial treatment, interior log maintenance is negligible. You may want to recoat the walls after several years to restore brilliance, but it shouldn't be necessary. Penetrating oil finishes, on the other hand, may need to be reapplied to condition the wood properly, as well as to maintain the finish.

Exterior maintenance should follow the recommendations of the exterior treatment manufacturer. Generally, this consists of a spray, roller, or brush application of treatment every three to five years. The appearance of the logs will provide some indication of the need for retreatment. If logs start graying or bleaching noticeably, or if prolonged exposure to wet weather leaves log surfaces looking damp, it's time to reapply. This generally happens first on exposed gable areas and areas with little roof overhang (a good reason for incorporating large overhangs into your design). If graying becomes too noticeable, the wood can be rebleached before retreating. This should restore a like-new appearance.

INSECTS

The problem with insects is that they don't pay attention. They ignore labels and textbook descriptions. That makes it hard to make generalizations and suggest specific treatments for specific problems.

Prevention is the best maintenance against insect harm. Building codes in many areas require various building features designed to discourage termite infestation. These include separating exposed wood from the ground by at least 8 inches of masonry or treated lumber. A termite pretreatment of poison in the excavation and around the base of the foundation is often required by lending institutions. A licensed pest control specialist must perform the work and will issue a certificate of completion that the bank will want (sometimes before issuing loan draws). In some areas termite shields are required or advised. Most builders that I have spoken with say that shields aren't worth the trouble it takes to install them, and most home owners don't want them because they are unsightly.

Termites are not the only potential threat to log homes, but you can take precautions to make your log home less desirable to other insect pests, as well. Do not store wood, wood scraps, and firewood within 50 feet of the house site. This helps discourage carpenter ants and centipedes. When removing trees from your building site, keep an eye out for signs of carpenter ant nests and other colonial insects such as bees. When their nest site is destroyed, they may take refuge in your nest site. Talk with an exterminator about how to deal with any nest that you uncover. Exterminators should be able to recommend a selection of treatments varying in cost, efficiency, and toxicity. Weigh their advice according to your conditions. Some people would rather swat bugs than loose a chemical cloud on their woodlands. It's a matter of your choice combined with your specific circumstance.

Insect problems tend to be site specific, or at least specific to an area. Of two log houses sitting side by side, made by the same manufacturer using the same wood, one house can be infested with powder post beetles, while the other has never yielded a single insect. When you think about it, this is exactly the sort of situation that confronts residents of conventional neighborhoods. The response is the same: call a pest control specialist.

As the number of log homes in the country has grown, companies specializing in insect treatments and pest control have begun to show more interest. Log home companies themselves, as they gain experience (the log home industry, as it is today, has only been around for about twenty years), have taken an interest in insect control. Many companies offer customers the option of having their logs chemically treated. Some companies fumigate logs before they are shipped.

This sort of preventive treatment is often the basis for insuring low maintenance during the life of the home. There are, however, some important considerations when chemicals are used to control insects. First of all, do they work? Specific chemicals have specific uses. As I said earlier, insect problems vary regionally and even locally. It's important to know that the treatment being offered is effective in your situation. Second, are there potential adverse effects to people and pets? Pesticides are, after all, poisons and over time may have adverse side effects. Since such effects may take years to appear, assurances that a new chemical is safe may not be enough.

Recently, borates, a class of chemicals with low toxicity, have gained favor in the log home community. According to studies, sodium borate is extremely effective against many kinds of insects without posing undue risk to you or your family dog or cat. Borate products are applied as a spray, or logs may be dipped into a solution. Chemical rods may be inserted into holes drilled periodically in the logs. These offer the advantage of a controlled release and dispersion throughout the log over time.

You may want to take up the question of preventive insect control with a local pest control company or with the county extension office of your state university.

By now you know that building a log home (or any home) is not a simple task. More than anything else, I hope you have developed a respect for careful planning, thorough research, and attention to detail. If so, you should be packing boxes and preparing to move into a log home well matched to your dream.

The Appendices contain a variety of forms that I have used to help organize and monitor quality on my log home projects. Each log home experience is unique, so you may have to make some modifications. If you choose to use the forms here, study them thoroughly before construction starts. Then, discipline yourself to stay with them. Consistency in planning and record-keeping will help control the amount of time required to keep your project running smoothly.

Resources and Supplies

PERIODICALS

Log Home Living
Home Buyer Publications
4451 Brookfield Corporate Drive, Suite 101
Chantilly, VA 22022
800/826-3893

Log Home Guide
Muir Publishing Company
164 Middle Creek Road
Cosby, TN 37722
800/345-5647

BOOKS

Burch, Monty. *Complete Guide to Building Your Own Log Home.* New York: Popular Science/Outdoor Life Books, 1984.
Heldmann, Carl. *How to Afford Your Own Log Home.* Chester, CT: Globe Pequot Press, 1984.
Lewis, Gaspar. *Carpentry.* New York: Sterling Publishing Company, 1984.

Mackie, B. Allen. *Building with Logs.* Log House Publishing Company, Ltd., 1979.

Petrocelly, Kenneth. *Build It Right — Supervising the Construction of Your New Home.* Blue Ridge Summit, PA: TAB Books, 1990.

Ramsey, Dan. *Building a Log Home from Scratch or Kit.* Blue Ridge Summit, PA: TAB Books, 1987.

Roskind, Robert. *Before You Build: A Pre-Construction Guide.* Berkeley, CA: Ten Speed Press, 1983.

Shepherd, James. *How to Contract the Building of Your New Home.* Williamsburg, VA: Shepherd Publishers, 1991.

Sherwood, Gerald, and Robert Stroh, eds. *Wood Frame House Construction.* Washington, DC: National Association of Home Builders, 1988.

SCHOOLS

Pat Wolfe Log Building School
R.R. 3
Ashton, Ontario KOA 1BO, Canada
613/253-0631

B. Allen Mackie School of Log Building
P.O. Box 2085
Vancouver, BC V6B 3T2, Canada
604/732-0774

Great Lakes School of Log Building
3544½ Grand Avenue
Minneapolis, MN 55408
612/822-5955

APPENDIX I.

BUILDING CONTRACT

Date: _____

1. The undersigned Purchaser: _____
authorizes OakRidge Log Homes, Inc., hereinafter referred to as Builder, to construct and deliver to said Purchaser a dwelling in accordance with the plans and specifications which are attached hereto and made part hereof, on that tract or parcel of land lying and being in Land:

Lot _____ of the _____
District of _____ County, _____ State, and being known as _____.

2. Except as proved in the paragraph dealing with allowances hereafter, the total cost to the Purchaser (including/excluding the cost of the land and additional arranged work) shall be:

_____ DOLLARS ($_____)

Said amount shall be paid as follows:

_____%	Initial payment to begin work	$_____
_____%	Foundation completed, first-floor joists and subfloor installed	$_____
_____%	Exterior log walls erected, exterior sheathing complete, basement floor/slabs complete	$_____
_____%	House under roof, windows, exterior doors installed, interior framing complete, shingles installed	$_____
_____%	Plumbing, heating, electrical wiring roughed in	$_____

_____%   Drywall complete, kitchen cabinets
            installed, plumbing fixtures
            installed, trim complete $_____

_____%   Driveway, final grading,
            liens for all labor and materials
            furnished, final Use and Occupancy
            Permit furnished $_____

3. Builder shall commence construction of the house in accordance with the attached plans and specifications after the construction loan has been closed. The construction of the house shall be completed within _____ days from the time that construction is commenced.

4. The improvements shall be deemed fully completed when approved by the mortgagee for loan purposes.

5. Builder agrees to give possession of the improvements to the Purchaser immediately after full payment to the Builder of any cash balance due. At that time, the Builder agrees to furnish the Purchaser a notarized affidavit that there are no outstanding liens for unpaid material or labor.

6. It is the understanding of the parties that the purchase price above described includes Builder's allowances for items hereinafter described. The amounts are listed at Builder's cost. It is the understanding of the parties that the purchase price shall be equal to the price described in Paragraph 2 above, provided the cost of the items below listed do not exceed the amounts shown. In the event that the Builder's cost exceeds any item hereinafter listed, the purchase price shall be increased by the amount above the allowance shown. The allowances are as follows:

Electrical service $_____
Electrical fixtures $_____
Cabinets/countertops/vanities $_____
Ceramic/marble $_____
Carpeting/vinyl floors $_____
Hardwood floors $_____
Septic system/sewer hookup $_____
Water line $_____
Sewer tap $_____
Water tap $_____
Well/pump $_____

Plumbing faucets/fixtures	$_____
Fireplace/woodstove	$_____
Appliances	$_____
Driveway/apron	$_____
Landscape/finish grade/seed	$_____

7. The cost of any alterations, additions, omissions, or deviations at the request of the Purchaser shall be added to or deducted from the agreed purchase price. No such changes shall be made unless stipulated in writing and signed by both parties hereto. Any changes that increase the cost of completing construction will be paid for by the Purchaser prior to commencement of work.

8. Builder shall provide to Purchaser, in writing, a twelve (12) month construction guarantee.

9. Every effort will be exerted on the part of the Builder to complete construction within the period described above. However, Builder assumes no responsibilities for delays occasioned by causes beyond his control, including the following:

a) Acts or default by Purchaser
b) Acts or default of any developer or contractor engaged in constructing or installing any streets or utilities
c) Adverse weather conditions
d) Damage caused by fire, storm, earthquake, or other casualty
e) Any form or act of God or force majeure
f) Government regulations, controls, procedures, or restrictions or moratoriums
g) Allocation of labor supplies or material by or under the authority of any government or government agency
h) Acts of suppliers of labor or materials
i) Acts of subcontractors or their employees
j) Purchaser's selections of items listed in Paragraph 6.

In the event of a delay of construction due to any of the causes just listed, the period of delay is to be added to the construction period allotted to the Builder.

10. In the event the Builder is unable to obtain the materials specified on the plans and specifications or the items shown in the selection sheets through reasonable sources of supply, the Builder shall have the right to substitute materials of similar pattern and design and substantially equivalent in quality. Builder reserves the right to make changes in plans and specifications solely for the purpose of mechanical installations, building code requirements, and normal architectural design improvements subsequent to the date of this agreement.

11. The movement of any household goods or other materials by the Purchaser into the house will not be permitted until the house has been completed and accepted by Purchaser and the sale price has been paid in full (closed).

12. In the event that on the settlement date scheduled in accordance with this agreement the home shall be otherwise habitable, but such items as landscaping, exterior concrete, driveways, final grading, and exterior painting cannot be completed by reason of weather conditions, settlement shall be consummated as scheduled as long as temporary access to the property shall be provided to the Builder. Builder agrees that such uncompleted work shall be completed as soon as weather conditions permit.

13. Any items determined to require corrective action as a result of the presettlement inspection conducted by Builder and Purchaser shall be completed as soon as is practical, but shall not be cause for delaying settlement if all other conditions of this contract have been fulfilled.

14. After completion of construction in accordance with this Agreement, settlement shall take place at the date and place selected by Builder on not less than ten (10) days notice to Purchaser. Upon payment by Purchaser of the balance due Builder and expenses of settlement and all proper fees and charges in connection with it, Builder shall convey the property to Purchaser by special warranty deed, title to be good and marketable, free from liens and encumbrances except as specified herein, and except for use and occupancy restrictions of public record and publicly recorded easements that may be observed by inspection of the property. Possession shall be given to Purchaser when Builder receives the balance of the purchase price; prior thereto Purchaser shall have no right to enter or occupy the property without written approval of Builder.

15. All closing costs and settlement charges (including but not limited to all conveyance fees, transfer taxes, recording fees), prepaid items (including but not limited to mortgage insurance premiums, prepaid fire and hazard insurance premiums, prepaid real estate taxes, and prepaid interest on the mortgage), and all other lender-required fees and charges shall be paid by Purchaser, except the following to be paid by Builder: _____

_____

_____

_____

Taxes and charges of every kind against the property that are or may be payable on an annual basis, including but not limited to city, county, and state valorem taxes; Homeowner's Association assessments; and other benefit charges and assessments for water, sewage, and other public improvements for the use thereof, shall be paid by Purchaser.

16. Default by Purchaser shall be deemed to have occurred upon Purchaser's failure to: a) make payments on or before the dates specified herein; b) on the date appointed to tender at settlement the amount called for herein and accept title, or c) to comply with other terms of this Agreement. In the event of Purchaser's default under this Agreement, Purchaser agrees that all sums of money paid hereunder prior to such default shall be retained by Builder as liquidated damages or, as an alternative, Builder may seek specific performance of this Agreement or any part thereof in a court of competent jurisdiction.

17. In the event Builder shall determine that, in good faith, and for reasons beyond its control, including any cause specified in Paragraph 9 and including any pending or declared governmental moratorium, that the home purchased hereunder cannot be completed and made available for occupancy prior to the time specified hereunder or within a reasonable time thereafter, or if Builder shall be unable to deliver good and marketable title to the property, this Agreement may be cancelled at the option of Builder upon ten (10) days' notice to Purchaser. In event of cancellation as provided for in this paragraph, Builder's responsibility shall be limited to the return of all monies paid hereunder by Purchaser and, upon such return, this Agreement shall be null and void and Builder shall be released from all obligations hereunder.

18. The property purchased hereunder and certain other property in the area hereof may be subject to certain restrictive covenants, agreements, easements, liens, charges, and restrictions as contained therein or provided for in certain instruments recorded among the land records of _____
_____ County, without limiting any provisions thereof. Purchaser accepts title to the property subject to the matters thus identified and to any other similar provisions contained in instruments

recorded among the appropriate land records prior to the transfer of title hereunder.

19. Purchaser's interest and obligation hereunder shall not be assignable without written consent of Builder.

20. The riders attached hereto and listed below are an integral part of this Agreement. _____

_____

_____

_____

_____

_____

_____

21. This contract shall inure to the benefit of, and shall be binding upon, the parties hereto, their heirs, successors, administrators, executors, and assigns.

22. This contract constitutes the sole and entire agreement between the parties hereto, and no modification of this contract shall be binding unless attached hereto and signed by all parties to the Agreement. No representation, promise, or inducement to this Agreement not included in this contract shall be binding to any party hereto.

The parties do set their hands and seals the day, month, and year first above written.

(Purchaser)	(date)
(Purchaser)	(date)
(OakRidge Log Homes, Inc., Builder)	(date)

APPENDIX II.

CONSTRUCTION STIPULATIONS

Special stipulations from sales contract dated _____ for the purchase of property located at _____.

1. The movement of any household goods or other materials into the house by Purchaser will not be permitted until house has been completed and accepted by Purchaser and the sale price has been paid in full (closed).

2. Every effort will be exerted on the part of the Seller to complete the construction of said house within the projected time; however, Seller assumes no liability for delays occasioned by causes beyond his control.

3. The costs of any alterations, additions, omissions, or deviations at the request of the Purchaser shall be added or deducted from the agreed sale price, but no such changes shall be made unless stipulated in writing and signed by all parties hereto. Any changes that increase the cost of completing construction will be paid for by the Purchaser prior to commencement of work.

4. Effective with the final acceptance and closing of transaction by the Purchaser, Seller will provide a ten-year home owner's warranty with exception of anything the Purchaser might have subcontracted on his own account.

5. The following standard features are included in the structure:

a) _____

b) _____

c) _____

d) _____

e) _____

f) _____

g) _____

h) _____

i) _____

6. House will be built per attached plans and specifications of materials marked Exhibits "C" and "D."

7. Seller stipulates that he will promptly correct any leakage or seepage of exterior surface water into basement of dwelling for a period of 12 months from the date this sale is closed. Seller will not be responsible for and will not correct any leakage or seepage caused by: a) breakage or bursting of water mains or pipes, b) any grading done by Purchaser that causes water to flow toward outside foundation wall, c) prolonged direction of water against the outside foundation wall from water spigot, sprinklers, hose, broken gutters clogged with leaves or pinestraw, or broken, bent, or clogged downspouts.

8. Decorator features such as carpet and appliances may be chosen by the Purchaser from the price range and samples offered by the suppliers chosen by the Seller. The following allowances are provided by the Seller. Purchaser may select a higher-cost item and if so selected, agrees to pay the difference in the allowance and the higher cost of items selected at time of selection:

a) Lighting fixtures including bulbs, chimes, tax
   $_____

b) Wall paper per single roll not installed
   $_____

c) Carpets per square yard including tax and installation
   $_____

d) Vinyl per square yard including tax and installation
   $_____

e) Foyer floor per square yard including tax and installation
   $_____

f) Landscaping allowance, if applicable
   $_____

g) Appliances (where applicable):
   (includes disposal, dishwasher, oven with hood, and stove)
   $_____

h) _____

    $_____

i) _____

    $_____

j) _____

    $_____

k) _____

    $_____

l) _____

    $_____

_____	_____
Purchaser	Seller

_____	_____
Purchaser	Seller

APPENDIX III.

CONTRACT SPECIFICATIONS

Specifications for:

1. GENERAL

1.1 Plans and architectural services required to obtain a building permit are to be provided by the ( ) Contractor ( ) Owner. The Owner shall provide to the Contractor sepias and three (3) sets of working drawings. Reproduction to produce additional sets is to be paid by the Contractor.

1.2 The Contractor shall obtain and pay for all permits and bonds required by the County for construction of the residence. These specifications assume that the Owner has a recorded building lot with approved septic or sewer connection and approved well or water connection. The Contractor includes an allowance of $_____ to drill, steel case, grout, and test the well. All engineering and survey work required to obtain or modify an approved building lot shall be paid for by the Owner.

1.3 Builder's risk insurance in an amount not less than the contract sum shall be maintained for the duration of the contract. In general, the policy shall provide all risk, damage, and liability coverage to protect both the Owner and the Contractor. The cost of the insurance shall be paid by the Contractor.

2. SITE PREPARATION

2.1 The Contractor shall stake the house location for final approval by the Owner. After the foundation is built, the Contractor shall provide a wall check survey. At the time of substantial completion the Contractor shall provide a final survey. The cost of survey work required to locate and/or permanently mark property corners/lines shall be the responsibility of the Owner.

2.2 The Owner and Contractor shall mutually agree upon which trees are to be cut and what area is to be cleared. Hardwood trees will be cut into 2' lengths and stored on-site up to a maximum of approximately five cords. Brush shall be chipped or hauled away. Tree stumps shall be hauled away or ground to a level at least 1' below finished soil surface. The Contractor is not responsible for the survival of uncut trees.

3. EXCAVATION

3.1 The Contractor shall excavate, provide standard backfill, and rough grade to obtain positive drainage away from the house. Final fine grading in preparation for seeding/sodding shall be part of the landscape allowance.

3.2 The following items are not included in the contract price and if required shall be billed as extras: a) nonstandard excavation necessitated by rock or underground springs, b) compaction of backfill, c) additional excavation or soil treatment caused by inadequate soil-bearing capacity, i.e. less than 2,500 pounds per square foot, d) importing or exporting of fill dirt or topsoil.

4. FOUNDATION

4.1 Footings shall be 2,500 psi concrete or as specified for foundation type. Size and reinforcement shall be in accordance with the plans.

4.2 Foundation wall to be _____

_____

_____

_____. Size and height of foundation walls as per plans.

4.3 Walls will be waterproofed with foundation tar, roller or spray applied.

4.4 Structural steel as per plans. Columns to be 3″ outside diameter (O.D.).

4.5 Sill plates shall be of pressure-treated lumber with sill sealer and anchor straps 6'-0″ on center (O.C.) minimum.

4.6   Foundation drain shall be 4″ perforated plastic pipe covered with clean gravel and 15# felt paper. Drain to grade outside basement foundation. See Plumbing Specifications (15) for more detail.

4.7   Basement floor to be 4″ concrete with 6 x 6 #10 wire mesh over 4″ gravel with 6 mil polyethylene vapor barrier.

4.8   Termite protection shall include a termite inspection and chlordane poisoning with a 2-year warranty.

4.9   Garage floor to be 4″ concrete with 6 x 6 #10 wire mesh and 6 mil polyethylene vapor barrier. At the garage door the concrete will be turned down to a depth of 18″ or a continuous footing will be provided. The garage floor will be sloped toward the doorway a minimum of 2″ for drainage. If fill dirt is required in the garage area, the slab will be reinforced with #4 rebar and grade beams as required by county codes.

5.   UTILITIES

5.1   Septic field and tank shall be installed in accordance with county requirements. System to be sized with/without garbage disposal. Septic allowance amount per allowance schedule. See Plumbing Specifications (15) for more detail.

5.2   The Owner shall dig, case, grout, and test a well in the approved location. The Contractor shall provide a connection from the well to the house. See Plumbing Specifications (15) for more detail.

5.3   The Contractor shall have electric and telephone service installed from the main lines to the house. The Contractor shall have the electric meter installed. The allowance amount for *(name of utility company)* to provide the service connection is $_____.

5.4   Main electric service is to be 200 amp.

5.5   The house shall be prewired for three (3) telephone outlets. Connections to the phone outlets shall be the responsibility of the Owner. Additional outlets will be installed for $30 each.

5.6   The house shall be prewired for two (2) TV antenna outlets, wires to terminate in attic for future antennas. Coaxial cable shall be used. At Owner's request, the wires can terminate for connection to cable TV service.

6. FRAMING

6.1 Floor joist size, spacing, and grade as per plans. If no grade is specified, #2KD SPF is standard.

6.2 Exterior walls to be 7½" x 5½" double tongue-and-groove oak or pine logs fastened with ⅜" x 8" plated lag screws every 2' O.C. Framed dormers or extensions as per plans. Exterior frame walls to be 2 x 6, 16" O.C. Studs to be KD-SPF. Headers over doors and windows to be two log courses with logs ending not less than 2' from edge of opening. Headers in framed bearing walls to be 2" x 12" unless shown otherwise.

6.3 Roof framing to be manufacturer's approved trusses and/or conventional framing with size, spacing, and grade as per plans.

6.4 Subflooring to be ¾" tongue-and-groove plywood glued and nailed to floor joists. Underlayment under ceramic floors to be ½" unless shown otherwise. ¼" masonite under vinyl floors unless shown otherwise.

6.5 Wall sheathing on framed exterior walls to be ½" plywood at corners and ½" Thermax elsewhere.

6.6 Roof sheathing to be ½" plywood installed with plyclips.

6.7 Exterior deck materials to be pressure-treated #2 Southern Yellow Pine (SYP) with #1 2 x 6 used for decking.

7. ROOFING

7.1 Shingles to be Owens-Corning Chaparral 25-Year Warranty Fiberglas or equivalent, installed over 15# felt with aluminum valley and step flashing.

7.2 Gutters and downspouts to be seamless aluminum in standard colors of brown, beige, or white. Splashblocks to be provided at terminal of downspout.

8. EXTERIOR SIDING

8.1 Exterior siding to be oak or pine log (see 6.2). Dormer, framed gables, and extensions sided with 2 x 6 tongue-and-groove pine log cabin siding. Soffit detail to be rough-sawn western red cedar.

9.   EXTERIOR WINDOWS AND DOORS

9.1   Exterior windows to be Crestline Smart-R double-hung terratone with screens and grills. Size and locations as per plans.

9.2   Exterior doors as per Door Schedule.

10.   INTERIOR WALLS

10.1   House will be hung with ½″ regular drywall on ceiling and walls. Shower and bathtub walls will be hung with ½″ water resistant board. Bottom 3 feet of all showers will be hung with Wonder Board or equivalent. Garage ceiling and walls adjacent to living space will be hung with ⅝″ firecode drywall. Remainder of walls in garage can be done for a charge of $48 per sheet. Contract includes drywall down basement stair wall on one side of studs. Remainder of basement is unfinished unless otherwise specified in plans or specifications.

10.2   Drywall will be nailed in a uniform manner with 1⅜″ or better drywall nails. Drywall adhesive will be applied to every stud and ceiling joist before application of drywall. Drywall may be fastened with screws instead of nails for an additional $500.

11.   FINISHING

11.1   Log walls will be sanded smooth prior to application of finish, and joints between logs will be caulked with a matching color latex caulk.

11.2   Drywall joints will be taped with three applications of drywall mud. Nail heads will be covered with drywall mud. All drywall will be sanded smooth with sandpaper. After primer coat of paint, defects in wall finish will be pointed up, sanded, and reprimed. Garage will receive only tape coat and block coat unless it is to be painted.

11.3   Exterior log walls and wood trim will receive two coats, wet-on-wet application, of a penetrating wood preservative. Preservative may be mixed with a light stain to minimize effects of UV light and irregularities in exterior log color.

11.4   Interior log walls will receive two coats of polyurethane varnish with a light sanding between coats.

11.5 Walls and drywall trim will receive two coats of Duron Pro-Kote paint, first coat consisting of spray-applied primer and second coat applied with brush or roller. All painted trim with exception of closet and linen shelves will be semigloss, and walls will be flat. Kitchen and bath areas can be done entirely in semigloss if Owner chooses. When custom colors are used, if they do not cover primer with one coat, a third coat will be applied at a charge of $.25 per square foot. Garage is not painted.

12. INTERIOR FINISH

12.1 Kitchen cabinets and countertops as per allowance schedule.

12.2 Vanities, mirrors, and shower/tub doors as per allowance schedule.

12.3 Stairs from first to second floor to be oak treads and risers, oak ballisters, oak handrail, and oak stringers. Stairs to basement to be #1 pine treads, risers, and stringers. Stair sizes, locations, and railing details as per plans.

12.4 Interior doors as per Door Schedule.

12.5 Closet shelving to be 1 x 12 or 1 x 16 Novaply, painted.

12.6 Interior trim as per Trim Schedule.

12.7 Every finished room will have baseboard. Carpet will be tucked to this where applicable. Where there is wood floor, ceramic floor, or vinyl floor a shoe mould will be used.

12.8 Closet trim in master bedroom as per plans. Other closets will receive one (1) pole and one (1) shelf at proper height with closet support used where span is greater than 4'. Linen closets to have five shelves — four 16" shelves and one 12" shelf on top.

13. ELECTRICAL

13.1 Standard wiring included in the contract is as follows:
   • Receptacles installed as per code
   • Lights in all bedroom closets
   • Two (2) double floods
   • Two (2) exterior receptacles
   • One (1) water heater connection
   • One (1) switch receptacle in each bedroom

- One (1) doorbell circuit
- Well pump as required
- Basement: two (2) receptacle and four (4) keyless
- Garage: one (1) receptacle, two (2) lights, garage-door-opener plugs
- Garage-door openers
- Appliances: refrigerator, dishwasher, disposal, range
- HVAC: as per plans
- Extra boxes to be installed at a cost of $30 each.
- Lighting fixtures to be charged against the lighting allowance include intercom, recessed light fixtures, bathroom fans, exterior lights, landscape lights, pole lights. Charge to hang extra light fixtures is $30 each.
- Three-way circuits for stairways, rooms with two entrances, front door, and one (1) additional door selected by Owner.

13.2 Unless otherwise noted in specifications, wiring for the following items is *not* included in the contract: Instant hot, separate microwave, freezer, trash compactor, exterior lights not attached to house (i.e. pole lamp or driveway lights), pool wiring, attic fan, whole-house fan, pull-down iron, whirlpool, steamers, four-way switches, dimmers, floor receptacles, under-cabinet lighting, assembly of track lights and/or chandeliers, sump pumps, condensate pumps, heat/light fans.

14. HVAC

14.1 Heating and air conditioning to be one (1) electric heat pump, Trane or equivalent. Size of system and number and location of supplies and returns to be determined by HVAC contractor.

14.2 System includes central humidifier.

14.3 Electric air cleaner and night set-back thermostat are available as options.

15. PLUMBING

15.1 Rough-in for basement bath. Future drain lines and vent stack connection will be provided at the finished concrete floor level for the bathroom configuration shown on the plans. If grade permits, drain lines will be connected to septic tank or sewer. If basement grade is above the septic tank or sewer elevation, then a sewage ejector pump will be installed at a cost of $250. The ejector pump, its installation at a future date, and electrical connection are *not* included in the contract. Water supply lines will be provided in the basement at a point that is accessible to the future bathroom. The contract does *not* include any stud walls, electrical outlets, or bathroom fans and/or fan connections.

15.2   The contract includes the installation of a Gould ½ hp two-wire, 220 volt or equivalent submersible well pump suitable for wells up to a depth of 240 feet. For deeper wells up to 340 feet, a Gould ¾ hp two-wire, 220 volt or equivalent submersible well pump will be used at an additional cost of $200. The pump will be connected to the house with a 1″ selflex or poly pipe. All pipe and connections for wells up to 100 feet from the house are included in the contract price. Additional distance from the house will be billed at a rate of $2.50 per foot.

15.3   The contract includes the installation of 1″ poly pipe from the public water house connection valve up to a distance of 100 feet. Additional distance from the house will be billed at a rate of $2.50 per foot. Water "tap" fees and the installation of the public water house connection valve, if required, are included as allowances.

15.4   Installation of the septic system is an allowance item. The system will be sized in accordance with county requirements for a three-bedroom house with garbage disposal. The type of system required may include drywell, leeching bed, deep trench, or shallow trenches and will be determined by county requirements. The labor and material items whose installed cost constitutes the septic field allowance include: septic tank, distribution box, pipe, stone, filter paper, block for drywells, lids, covers, and cleanouts, trenching, excavation, and backfill. Importing or exporting fill is *not* included in the cost.

15.5   The contract includes installation and connection of a 4″ PVC pipe from the house to an existing sanitary sewer house connection whose location is marked by a visible wood marker. The maximum distance from the house to the sewer connection included in the contract price is 50 feet. Additional cost for longer distances is $10 per foot. The maximum depth of the sewer at any point between the house and the connection is assumed to be 12 feet. Additional depth will be billed as a change order. Sewer "tap" fees, if required, are included as allowances.

15.6   The contract includes (<u>brand name</u>) or equivalent water heater(s). If two heaters are provided, water distribution will be balanced between supply points in the house. If oil or gas heaters are used, they will be vented according to manufacturer's requirements.

15.7   A _____ or equivalent model well pump pressure tank will be installed.

15.8   Water supply lines will be CPVC plastic or schedule M copper. The size of the lines, unless otherwise specified, will be determined by the plumbing subcontractor. There will be one main supply cut-off valve accessible in the basement and individual cut-off valves to other fixtures. There will be _____ outside frost-free hose bibs located as per the plans. If not specified on the plans, they will be distributed, one each, to the front, rear, and garage parking areas.

15.9   Waste and vent lines will be PVC. Their location and size will be determined by the plumbing subcontractor. Wherever possible, vent stacks will be located on the rear of the roof. Options available to the Owner are: a) main vertical drain located in a 6″ wall with insulation to reduce noise ($30), and b) paint vent stacks above roof to match roof color approximately ($30).

15.10 If the house has a walk-out basement, a 1½″ PVC drain line will be run to daylight. This line will receive discharges of HVAC condensate, water heater emergency overflow, and humidifier emergency overflow. If there is no walk-out basement, then a sump pit with pump will be provided to receive the above mentioned discharges. Options available to Owner are: HVAC condensation pump for direct discharge to outside ($150 each).

15.11 If the house has a walk-out basement, a 4″ diameter perforated plastic drain pipe, discharging to daylight, will be installed around the outside of the foundation. The pipe will be covered with ½″ diameter gravel and 15# red rosin paper in accordance with county requirements. If there is no walk-out, the 4″ pipe will drain to a plastic pump pit fitted with a submersible sump pump. All basement area-way drains will drain to this interior sump pit. Options available to Owner for clay conditions or low-lying areas are: outside drain to daylight *and* inside drain to sump pit with pump ($275).

15.12 **Plumbing Fixtures**

Kitchen
Sink
Sink faucet
Instant hot
Disposal*
Icemaker*
Island sink
Island sink faucet
Other

Powder Room
Toilet
Toilet fixtures
Sink
Sink faucets
Other

Basement Bath
Tub
Tub faucets
Shower
Shower faucets
Sink
Sink faucets
Toilet
Toilet fixtures

First-Floor Bath
Tub
Tub faucets
Shower
Shower faucets

Sink

Sink faucets

Toilet

Toilet fixtures

Second-Floor Bath

Tub

Tub faucets

Shower

Shower faucets

Sink

Sink faucets

Toilet

Toilet fixtures

Master Bath

Tub

Tub faucets

Shower

Shower faucets

Sink

Sink faucets

Toilet

Toilet fixtures

Bidet

Steamer

*Icemaker, disposal, and ceramic tile are allowance items.
**Tubs and toilets are in standard colors.

16.   INSULATION

16.1   Vaulted ceiling insulation to be R-19 or R-30 depending on depth of
rafter with baffles, if required. Crawlspace insulation to be R-19.
Exposed above-grade basement walls at walk-out to be insulated
with R-8 in accordance with county requirements. Ceiling insulation to be R-30.

17. FLOOR COVERINGS

17.1 Carpet per allowance schedule.

17.2 Ceramic tile included as allowance.

17.3 Vinyl included as allowance.

17.4 Finish floors based on Floor Schedule.

18. LANDSCAPING

18.1 Final finish grade, seeding, sodding, fountains, and plantings will be installed and included as allowances.

18.2 Driveway will be twelve (12) feet wide with a 20' x 35' parking apron consisting of 4″ to 6″ of crusher run stone. Location as per revised site plan.

18.3 Walkways will be up to 30 pieces of 18″ x 24″ flagstone, set and leveled.

19. APPLIANCES

19.1 Appliances per allowance schedule.

20. HARDWARE

20.1 Kitchen and vanity hardware installed as part of allowance.

20.2 Interior locks to be _____
     style, _____ finish.

20.3 Exterior locks to be _____.

20.4 Exterior deadbolts to be _____.

20.5 Front door allowance for standard bore lock $_____.

20.6 Each bathroom will receive one (1) _____ toilet paper holder and two (2) _____ 24″ towel bars.

20.7 Polished brass doorstops, hinges, and pulls on interior doors.

APPENDIX IV.

PRECONSTRUCTION CHECKLIST

Project name: _____ Start date: _____
Project #: _____

Item	Date	Remarks
☐ Perk test	_____	_____
☐ Survey complete	_____	_____
☐ Loan application filed	_____	_____
☐ Loan application approved	_____	_____
☐ Bids sent	_____	_____
☐ Bids received	_____	_____
☐ Estimate complete	_____	_____
☐ Building contract signed	_____	_____
☐ Log kit contract signed	_____	_____
☐ Site plan complete	_____	_____
☐ Blueprints complete	_____	_____
☐ Building permit filed	_____	_____
☐ Building permit approved	_____	_____
☐ Subcontractors notified	_____	_____
☐ Log delivery scheduled	_____	_____
☐ Utilities scheduled	_____	_____
☐ Portable toilet scheduled	_____	_____
☐ Storage van scheduled	_____	_____
☐ Builder's risk insurance filed	_____	_____

Appendix V.

Requests for Quotations

PLUMBING

Project name: _____ Start date: _____

Address: _____

_____

| City | County | State | Zip |

Directions: _____

_____

_____

_____

☐ Well depth: _____ Distance to house: _____

☐ Public water hook-up allowance: _____

☐ Septic system distance to house: _____

☐ Public Sewer hook-up allowance: _____

☐ Heating system

    ☐ Hot-water baseboard   ☐ Natural gas

    ☐ Propane   ☐ Oil

Please provide a firm bid to perform the following work. Unless otherwise specified, include all materials, labor, taxes, and permit fees for the items checked. Please include manufacturer, model numbers, and color/styles (if not stated) where requested. Thank you for your prompt attention.

☐ Permit(s)

☐ Well pump/tank (specify make/model/size)

☐ Sump pump

☐ Basement floor drain

☐ Basement bath   ☐ rough-in   ☐ finished
☐ Sewage ejector pump
☐ Heating system (as specified above)
☐ Hot-water heater(s): Qty. _____   Size _____
☐ Kitchen sink: Type _____   Size _____
☐ Kitchen sink faucet: Type _____ ☐ Spray
☐ Disposal (specify make/model/size)
☐ Instant hot
☐ Ice maker
☐ Bath fixtures

_____   Toilets: Type _____   Color _____
_____   Lav sinks: Type _____   Color _____
_____   Showers: Type _____   Color _____
_____   Tubs: Type _____   Color _____
_____   Shwr/tub: Type _____   Color _____
_____   Shwr/tub doors: Type _____   Color _____
_____   Lav faucets: Type _____   Color _____
_____   Shwr diverters: Type _____   Color _____
_____   Tub faucets: Type _____   Color _____

☐ Master bath

_____   Toilet: Type _____   Color _____
_____   Lav sinks: Type _____   Color _____
_____   Shower: Type _____   Color _____
_____   Tub: Type _____   Color _____
_____   Shwr/tub: Type _____   Color _____
_____   Shwr/tub doors: Type _____   Color _____
_____   Lav faucets: Type _____   Color _____
_____   Shwr diverters: Type _____   Color _____
_____   Tub faucets: Type _____   Color _____
_____   Whirlpool: Type _____   Color _____

☐ Washing machine hook-up: Location _____
☐ Other _____
_____
_____

Plumber is responsible for scheduling and completing all necessary inspections.

All packaging, scrap, and trash from work to be removed from site, owner's manuals and warranty cards given to Builder.

Plumber must provide insurance certificate upon approval of bid.

Advance notice required to schedule work is about _____ days.

Authorized Signature	Date

ELECTRICAL

Please provide a firm bid to perform the following work. Unless otherwise specified, include all materials, labor, taxes, and permit fees for the items checked. Please include manufacturer, model numbers, and color/styles (if not stated) where requested. Thank you for your prompt attention.

Project name: _____ Start date: _____

Address: _____

_____

_____

City   County   State   Zip

Directions: _____

_____

_____

_____

☐ Service panel: _____ amp.   ☐ Temporary pole/service panel

☐ Heating system

  ☐ Electric baseboard   ☐ Heat pump(s) _____ amp.

  ☐ Forced-air electric   ☐ Gas/oil

  ☐ Outside A/C disconnect _____ amp.

☐ Permit(s)

☐ _____ Total receptacles (includes those listed below)

☐ _____ Switched outlets

☐ _____ Brass floor outlets

☐ _____ Weatherproof outlets

☐ _____ Smoke detectors

☐ _____ Phone outlets

☐ _____ Cable TV outlets   ☐ attic termination for antenna

☐ _____ Floodlights   ☐ Motion sensor(s) _____

☐ _____ Paddle fan/lights   ☐ Assemble and hang

☐ _____ Lights in beams

☐ _____ Outlets in log walls

☐ _____ Bath light/fan units _____ or specify

☐ _____ Recessed lights

☐ _____ Pull chain/keyless light fixtures

☐ _____ Heat lamps

☐ _____ Closet fluorescent with wall switch

☐ Intercom

☐ Well pump/tank   ☐ Sump pump   ☐ Sewage ejector pump

☐ Heating system (as specified above)

☐ Hot-water heater(s): Qty. _____   Size _____

☐ Disposal   ☐ Instant hot   ☐ Ice maker

☐ Range/oven   ☐ Microwave oven   ☐ Cook-top

☐ Whirlpool

☐ Washing machine   ☐ Dryer   ☐ Dryer vent

☐ Door chimes: ☐ Front door   ☐ Back door

☐ Price to hang light fixtures $_____ per fixture

☐ Other _____
_____
_____
_____

Electrician is responsible for scheduling and completing all necessary inspections.

Wire chases, holes, and receptacles in logs and beams will be drilled or routed by carpenters; wire will be laid in beam chases by builder according to electrician's specifications; wire to be provided by electrician.

All packaging, scrap, and trash from work to be removed from site, owner's manuals and warranty cards given to Builder.

Electrician must provide insurance certificate upon approval of bid.

Advance notice required to schedule work is about _____ days.

Authorized Signature	Date

SITEWORK AND FOUNDATION

Please provide a firm bid to perform the following work. Unless otherwise specified, include all materials, labor, taxes, and permit fees for the items checked. Please include manufacturer, model numbers, and color/styles (if not stated) where requested. Thank you for your prompt attention.

Project name: _____ Start date: _____

Address: _____

_____

_____

             City   County   State   Zip

Directions: _____

_____

_____

_____

☐ Site preparation

     ☐ Clear house area   ☐ Clear septic area

     ☐ Remove stumps from property

     ☐ Dispose of stumps on property by: _____

     ☐ Cut fallen trees to 4' lengths and stack

     ☐ Cut fallen trees to 16″ lengths and stack

☐ Entrance road

     ☐ Clear and grade entrance road

     ☐ Dispose of trees and stumps as shown on site plan

     ☐ Gravel entrance road with base of 2″ stone topped with 4″ crushed rock, packed

     ☐ Install culverts or drainage pipes as shown on site plan

     ☐ Install asphalt/concrete apron at entrance from public road

☐ Excavation/grading

   ☐ Excavate basement area, topsoil and fill piled separately

   ☐ Install silt/sedimentation control as shown on attachment

   ☐ Dig _____ linear feet of footings (Size: ____ x ____)

   ☐ Dig _____ linear feet of frost footings (Size: ____ x ____)

   ☐ Dig _____ fireplace footing (Size: ____ x ____)

   ☐ Dig _____ support piers (Size: ____ x ____)

   ☐ Backfill/rough grade foundation

   ☐ Finish grade

☐ Install septic system

   ☐ Permit: Tank size _____ gal.   System type _____

☐ Foundation

   ☐ Pour piers and footers

   ☐ Foundation walls

      ☐ Poured concrete: _____ thick x _____ height

Note: Unless otherwise specified, estimates for poured concrete wall should include 12″ anchor bolts placed on 6′ centers.

      ☐ Masonry block: _____ thick x _____ courses

Note: Unless otherwise specified, estimates for masonry block wall should include parging with portland cement, durawall metal reinforcement every three courses, FHA Cap Block for top course, 12″ anchor bolts placed on 6′ centers, columns reinforced with #2 rebar and poured solid every 6′.

      ☐ Masonry/stone fireplace as shown on attached plan

      ☐ Masonry/stone veneer on foundation/fireplace

         ☐ Stone   ☐ Cultured stone

      ☐ Apply waterproofing

      ☐ Install drain tile, filter paper

      ☐ Install radon loop

      ☐ Pour foundation/garage slab reinforced with 4″ gravel base, 6 mil poly vapor barrier, 6″ wire mesh, #2 rebar on 4′ centers

☐ Contractor   ☐ Builder   is responsible for scheduling and completing all necessary inspections.

All packaging, scrap, and trash from work to be removed from site, owner's manuals and warranty cards given to Builder.

Contractor must provide insurance certificate upon approval of bid.

Advance notice required to schedule work is about _____ days.

_____   _____
      Authorized Signature   Date

Appendix VI.

Technical Specifications

DRYWALL SPECIFICATIONS

Name: _____   Start date: _____

Address: _____   Contact: _____

_____   Phone: _____

- Bid is to provide all material, labor, and equipment to perform complete job per specifications of plans. This includes:
    - Stock drywall
    - Hang
    - Finish joints
    - Point up (retouch) joints after prime coat
    - Provide own ladders and scaffolding

- Apply ½″ gypsum board (drywall), double nailed at top. Four nails per stud, on interior partition walls and indicated ceiling areas only.

- Use ⅝″ fire-rated drywall for all common walls and ceilings between house and attached garage.

- All ceilings to be fastened using drywall screws.

- All joints to be taped with three (3) separate coats of joint compound, each sanded smooth to the touch.

- All outside corners to be reinforced with metal corner beads.

- All inside corners to be reinforced with joint tape.

- All necessary electrical outlet, switch, and fixture cutouts to be made.

- All necessary HVAC ductwork cutouts to be made.

- Wall adhesive to be applied to all studs prior to drywall application.

- Moisture-resistant gypsum to be used along all wall areas around shower stalls and bathtubs.

- Use drywall stilts at your own risk.

- Removal from site of all trash and debris resulting from work.

- Furnish certificate of insurance upon approval of bid.

ELECTRICAL SPECIFICATIONS

Name: _____   Start date: _____

Address: _____   Contact: _____

_____   Phone: _____

- Bid is to perform complete electrical wiring per attached drawing(s).

- Service to be _____ amp.

- Install electric baseboard heat per room size requirements.

- Include all supplies except lighting fixtures and appliances. Bid should include installation of light fixtures and outlets or hook-ups for appliances.

- Include temporary electric pole and temporary hook-up to power line.

- Outlet and switch openings in log walls will be routed by builder. Wiring in log walls to be done as recommended in manufacturer's construction manual (pages attached).

- All work and materials to meet or exceed all requirements of the National Electrical Code unless otherwise specified.

- Natural earth ground to be used.

- Remove from site all trash and debris resulting from work.

- Furnish certificate of insurance upon approval of bid.

PLUMBING SPECIFICATIONS

Name: _____   Start date: _____

Address: _____   Contact: _____

_____   Phone: _____

- Bid is to perform complete plumbing job for dwelling as described on attached drawing(s).

- Bid is to include all material and labor including all fixtures shown on attached drawing(s).

- All materials and work shall meet or exceed all requirements of the local plumbing code. In absence of local code, apply national code standards (BOCA or equivalent).

- No plumbing, draining, and venting to be covered, concealed, or put into use until tested, inspected, and approved by local inspectors.

- Plumbing inspection to be scheduled by plumber. Builder to be notified in advance of scheduled inspection.

- All plumbing lines shall be supported so as to insure proper alignment and prevent sagging.

- Install approved pressure-reduction valve at water service pipe.

- Plumb water supply so as to eliminate completely water hammer.

- Water pipes to be of adequate dimension to supply all necessary fixtures simultaneously.

- Furnish certificate of insurance upon acceptance of bid.

EXCAVATION SPECIFICATIONS

Name: _____   Start date: _____
Address: _____   Contact: _____
_____   Phone: _____

- Excavator to perform all necessary excavation and grading as indicated below and on attached drawing(s) and provide all necessary equipment to complete the job.

- Bid to include all equipment and equipment drag time for two trips.

- Bid to include necessary chain saw work.

- Trees to be saved will be marked with red ribbon. Trees to be saved are to remain unscarred and otherwise undamaged.

- Foundation to be dug to depth indicated at foundation corners. Foundation hole is to be three feet wider than foundation for ease of access. Foundation base not to be dug too deeply in order to pour footings and basement floor on stable base.

- Foundation floor to be dug smooth within two inches of level.

- Provide cost per truckload to import or remove dirt.

- Driveway area to be cleared of topsoil and cut to proper level. Topsoil to be piled separately.

- All stumps and trees to be dug up and hauled away. Hardwoods to be cut up to fit in fireplace and piled on site where directed.

- Grade level will be established by builder. Survey of the lot showing location of the dwelling and all underground utilities will be furnished. Underground utilities will be marked on site.

- Finish grade to provide positive drainage from house.

- Backfill foundation walls after waterproofing, gravel, and drain tiles have been installed and inspected.

- Builder to be notified prior to clearing and excavation. Builder must be on site at start of operation.

- Furnish certificate of insurance upon approval of bid.

FOUNDATION SPECIFICATIONS

Name: _____   Start date: _____
Address: _____   Contact: _____
_____   Phone: _____

- Bid to include locating, staking, and digging footings.

- Subcontractor will arrange for footing inspection.

- Pour concrete for footings as per applicable building code requirements.

- Lay out foundation and install forms for poured concrete or masonry block as per attached drawing(s).

- All masonry, concrete, form, and finish work is to conform to the local building code.

- Concrete will not be poured if precipitation is likely unless otherwise instructed.

- All masonry, form, finishing, and concrete work must be within one fourth of one inch of level.

- Install waterproofing, parging, or insulation board.

- Install drain tile around footings.

- Concrete is to be air-entrained ASTM Type I (General Purpose), 3,000 psi after 28 days.

- Concrete is to be delivered to site and poured into forms in accordance with generally accepted standards.

- Washed gravel and concrete silica sand to be used.

- Each concrete pour to be done without interruption. No more than a half hour between loads of cement to prevent poor bondage and seams.

- Remove from site all debris and trash that results from work.

- Furnish certificate of insurance upon approval of bid.

HVAC SPECIFICATIONS

Name: _____   Start date: _____
Address: _____   Contact: _____
_____   Phone: _____

- Bid is to install heating ventilation and air conditioning system (HVAC) in dwelling shown on attached drawing(s).

- Install electric heat pump(s) or _____ of appropriate size with air conditioner and humidifier, including unit, all ductwork, outside compressor unit.

- Install drain line to sump or exterior for AC drainage.

- Install downdraft line for range/cooktop, if applicable.

- Install dryer vent and vent lines for bathroom fans.

- All work and materials to meet or exceed requirements of local building code. In absence of local code requirements, apply national code requirements (BOCA or equivalent).

- Bid to include any necessary inspections. Notify builder prior to inspection.

- Furnish certificate of insurance upon approval of bid.

PAINTING SPECIFICATIONS

Name: _____   Start date: _____

Address: _____   Contact: _____

_____   Phone: _____

- Include all materials, labor, and tools.

- Primer coat and one finish coat to be applied to all drywall surfaces including closet interiors.

- Walls to be touch-sanded after primer has dried.

- Flat latex to be used on drywall; polyurethane varnish to be used on all stained or natural finished trim work.

- Log walls, exposed beams, and trim to be coated with polyurethane varnish.

- Ceiling white to be used on all ceilings.

- All trim joints to be caulked and sanded before painting.

- All paint to be applied evenly on all areas to be painted.

- Window panes to be cleaned by painter.

- Clean up after job, including removal from site of all trash and debris resulting from work.

- Excess paint to remain on site when job is completed.

- All exterior log walls and trim to be coated with spray-applied wood preservative supplied by builder.

- Furnish certificate of insurance upon approval of bid.

APPENDIX VII.

DAILY SITE REPORT

Project name: _____   Date: _____

Site manager: _____

Weather: ☐ Fair   ☐ Overcast   ☐ Rain   ☐ Snow

   Temperature: ☐ <32   ☐ 32–50   ☐ 50–80   ☐ >80

   Wind: ☐ Still   ☐ Light   ☐ High

   Remarks: _____

Workers on site:

____ Foreman/crew leader	____ Bricklayers	____ Masons
____ Foundation	____ Concrete finish	____ Excavator
____ Plumbers	____ Electricians	____ HVAC
____ Framers	____ Trim carpenters	____ Roofers
____ Insulators	____ Drywallers	____ Painters
____ Cabinet installers	____ Floor finishers	____ Other
____ Ceramic tile	____ Floor coverings	____ Other

Remarks: _____

_____

_____

Equipment on job: _____

_____

_____

Work completed: _____

_____

_____

_____

Work in progress: _____

_____

_____

_____

Equipment/material needs: _____

_____

_____

_____

Visitors/conversations/problems noted: _____

_____

_____

Condition of site: ☐ Satisfactory   ☐ Unsatisfactory

Explain: _____

_____
Signature

APPENDIX VIII. Cost Estimate Checklist

	A	B	C	D	E	F	G	H	I
			Material		Labor		Subcontract		
1	Project name:		Date:						
2	Contact:		Phone:						
3									
4		Quantity	Unit Price	Total Mat'l	Unit Price	Total Labor	Unit Price	Total	Total
5	**Financing Costs**								
6	Fees (FHA, VA, other)								
7	Construction loan interest								
8	Discount & closing costs								
9	Other								
10	Other								
11									
12	**Land**								
13	Raw land / lot								
14	Other								
15	Other								
16									
17	**Permits & Preliminary Costs**								
18	Survey/plot plans								

Cost Estimate Checklist (continued)

	A	B	C	D	E	F	G	H	I
19	Building permits								
20	Temporary electrical								
21	Builder's risk insurance								
22	Other								
23	Other								
24									
25	**Site Preparation**								
26	Clear site								
27	Rough grade								
28	Excavation								
29	Road/driveway								
30	Other								
31	Other								
32									
33	**Foundations/Piers/Slabs**								
34	Footings								
35	Walls/piers								
36	Slab basement/garage								
37	Sill plates								
38	Floor joists & decking								

Cost Estimate Checklist (continued)

	A	B	C	D	E	F	G	H	I
39	Vents								
40	Insulation								
41	Termite protection								
42	Drains								
43	Waterproofing								
44	Backfill								
45	Other								
46	Other								
47									
48	**Log Home Package**								
49	Basic package								
50	Garage options								
51	Dormer options								
52	Porch options								
53	Other options								
54	Windows								
55	Exterior doors								
56	Interior doors								
57	Storm windows & screens								
58	Storm doors								

Cost Estimate Checklist (continued)

	A	B	C	D	E	F	G	H	I
59	Freight								
60	Package sales tax								
61	Erection of log home package								
62	Unloading / forklift rental/labor*								
63	Other								
64	Other								
65									
66	**Roof & Exterior Trim**								
67	Roof materials/shingles								
68	Roof sheeting								
69	Roof insulation & paper								
70	Roof system, trusses & rafters								
71	Dormer materials								
72	Gable end materials								
73	Second-floor subflooring								
74	Second-floor girder & joists								
75	Flashing for windows & doors								
76	Window & door trim								
77	Other								
78	Other								

Cost Estimate Checklist (continued)

	A	B	C	D	E	F	G	H	I
79									
80	**Brick & Masonry***								
81	Brick								
82	Stucco								
83	Driveway								
84	Walks								
85	Porch								
86	Steps								
87	Patio								
88	Other								
89	Other								
90									
91	**Gutters & Downspouts**								
92	Gutters								
93	Downspouts								
94	Splash blocks								
95	Other								
96									
97	**Fireplace***								
98	Fireplaces								

Cost Estimate Checklist (continued)

	A	B	C	D	E	F	G	H	I
99	Chimneys								
100	Facing materials								
101	Mantels								
102	Other								
103									
104	**Plumbing**								
105	Kitchen sink*								
106	Vanities*								
107	Toilets*								
108	Bathtubs*								
109	Showers*								
110	Laundry tray								
111	Drain, waste & vent								
112	Water piping								
113	Water heater								
114	Water softener								
115	Gas piping								
116	Bath accessories								
117	Other								
118									

Cost Estimate Checklist (continued)

	A	B	C	D	E	F	G	H	I
119	**Electrical**								
120	Panel box								
121	Wiring								
122	Outlets								
123	Light fixtures*								
124	Smoke detectors								
125	Doorbells*								
126	Other								
127	Other								
128									
129	**HVAC**								
130	Heating system								
131	Cooling system								
132	Thermostat								
133	Attic fan								
134	Kitchen exhaust								
135	Bath exhausts								
136	Other								
137	Other								
138									

Cost Estimate Checklist (continued)

	A	B	C	D	E	F	G	H	I
139	**Insulation**								
140	Ceiling								
141	Floor								
142	Wall								
143	Duct								
144	Pipes								
145	Other								
146									
147	**Interior Wall Finish**								
148	Gypsum board								
149	Paneling								
150	Interior partitions								
151	Other								
152	Other								
153									
154	**Interior Carpentry**								
155	Stairs (basement)								
156	Stairs (finish)								
157	Passage doors								
158	Closet doors								

Cost Estimate Checklist (continued)

	A	B	C	D	E	F	G	H	I
159	Closet shelves								
160	Door hardware*								
161	Base mold								
162	Window trim								
163	Other trim								
164	Other								
165	Other								
166									
167	**Kitchen Cabinets***								
168	Cabinets								
169	Countertops								
170	Other								
171	Other								
172									
173	**Ceramic Tile***								
174	Bathtubs								
175	Showers								
176	Wainscot								
177	Floors								
178	Countertops								

Cost Estimate Checklist (continued)

	A	B	C	D	E	F	G	H	I
179	Other								
180									
181	**Interior Painting**								
182	Ceilings								
183	Walls								
184	Wallpaper								
185	Trim								
186	Floor								
187	Other								
188									
189	**Flooring***								
190	Wood								
191	Resilient/tile								
192	Carpet								
193	Pad								
194	Stone								
195	Other								
196	Other								
197									
198	**Appliances***								

Cost Estimate Checklist (continued)

	A	B	C	D	E	F	G	H	I
199	Refrigerator								
200	Oven/range								
201	Dishwasher								
202	Disposal								
203	Trash compactor								
204	Microwave								
205	Washer								
206	Dryer								
207	Other								
208									
209	**Utility Hook-up***								
210	Water well								
211	Community water								
212	Sewer/septic								
213	Electrical								
214	Phone								
215	Cable TV								
216	Other								
217	Other								
218									

Cost Estimate Checklist (continued)

	A	B	C	D	E	F	G	H	I
219	**Landscaping***								
220	Topsoil								
221	Seeding/sod								
222	Plants								
223	Other								
224									
225	**Clean-up**								
226	Trash removal								
227	Other								
228	**Totals**								

*These items may be included as allowances.

Appendix IX.

Dream Features/Scope of Project

Exterior Appearance

Log Styles
  milled
  hand-hewn with chinking between logs

Corners
  crossed corner (butt & pass)
  dovetailed

Roof System
  conventional rafter or truss roof
  heavy timber roof with exposed timber rafters

Roof Coverings
  conventional shingles
  cedar shakes
  metal
  slate

Walkout basement

Garage (attached, separate, basement)

Interior Features

For these features you will need to consider individual rooms or areas of the house.

Second Floor System
  Conventional Second Floor System (conventional joists covered with
    ceiling material.
  Exposed beams

Ceilings and Ceiling Coverings
Cathedral Ceilings (where?)
Wood, Tongue-and-Groove
Drywall

Floor Coverings
Hardwood flooring
Carpet
Slate
Vinyl sheet or tile
Other

Interior Partition Wall Coverings
Wood, Tongue-and-Groove
Drywall
Masonry
Other

Kitchen
Design (open to dining area, separate, breakfast nook?)
Cabinetry, (type and style)
Microwave
Dishwasher
Disposal
Sink Type
Cooktop or indoor grill
trash compactor
instant hot water
refrigerator/freezer
countertops
   Formica® or similar
   ceramic tile
   wood (butcher block)
   marble
   granite
   Corian® or similar

Baths
Tubs, Showers or combination
Style and color of fixtures
Style and type of vanity cabinet

Medicine Chests, Mirrors
    (mounted on wall or recessed in)
    whirlpool tub (where? _____)
    hot tub (where? _____)
    sauna/steam room (where? _____)

Decks, porches and balconies
    Railing Style

Other
    sunken living room
    greenhouse, sunroom
    pool, lap pool
    workshop
    darkroom
    studio/craft room
    fireplace
    woodstove

Appendix X.

Land Locator Checklist

Property Location: _____

Owner/Agent Name and Address: _____

Asking Price: _____

Property Taxes: _____

Recent Survey? _____

Recorded deed?   Deed #_____

Well: Yes   No   _____

Septic: Yes   No   _____

   If no septic, does land have successful perk test? _____

Wooded Lot? _____

Sloped for Walk-out basement? _____

Driveway/entrance road present? _____

School District: _____

Nearest Fire Station: _____

Nearest Emergency Medical Service: _____

Shopping: _____

All-weather road access: _____

Will easements or rights-of-way be required for entry? _____

APPENDIX XI. SAMPLE SCHEDULE

Project name: _____ State date: _____ Est. Completion Date: _____

ITEM	WEEK 1				
	MONDAY	TUESDAY	WEDNESDAY	THURSDAY	FRIDAY
Preconstruction	Post Building Permit				
Road/Driveway	Layout Road	Clear Road	Gravel Road	Drill Well	Test Well
Well					
Septic System					
Site Preparation			Clear Lot	Cut, Stack Wood	Layout Foundation
Excavation					
Foundation					
Termite Treatment					
Slab					
Framing					
Roofing					
HVAC					
Plumbing					
Electric					
Insulation					
Drywall					
Painting					
Trim					
Appliances					
Cabinets					
Floor Coverings					
Gutters					
Landscaping					

SAMPLE SCHEDULE

ITEM	WEEK 2				
	MONDAY	TUESDAY	WEDNESDAY	THURSDAY	FRIDAY
Preconstruction					
Road/Driveway					
Well					
Septic System					
Site Preparation					
Excavation	Dig Foundation	Dig Footings, Inspect			
Foundation			Pour Footings		
Termite Treatment					Sched. Pre-Treatment
Slab				Schedule Slab	Order Subfloor
Framing				Erect Walls	Erect Walls
Roofing					
HVAC					
Plumbing					
Electric					
Insulation					
Drywall					
Painting					
Trim					
Appliances					
Cabinets					
Floor Coverings					
Gutters					
Landscaping					

SAMPLE SCHEDULE

ITEM	WEEK 3				
	MONDAY	TUESDAY	WEDNESDAY	THURSDAY	FRIDAY
Preconstruction					
Road/Driveway					
Well					
Septic System					
Site Preparation					
Excavation					
Foundation	Erect Foundation	Erect Foundation	Waterproof Draintile		
Termite Treatment				Termite Treatment	
Slab			Set Subfloor	Prep Slab	Pour Slab
Framing					
Roofing					
HVAC					
Plumbing					
Electric					
Insulation					
Drywall					
Painting					
Trim					
Appliances					
Cabinets					
Floor Coverings					
Gutters					
Landscaping					

Sample Schedule

WEEK 4

ITEM	MONDAY	TUESDAY	WEDNESDAY	THURSDAY	FRIDAY
Preconstruction					
Road/Driveway					
Well					
Septic System					
Site Preparation					
Excavation	Backfill				
Foundation					
Termite Treatment					
Slab					
Framing		Deliver Logs	Set Logs	Set Logs	Set Logs
Roofing					
HVAC					
Plumbing	Select Fixtures				
Electric					
Insulation					
Drywall					
Painting					
Trim					
Appliances					
Cabinets					
Floor Coverings					
Gutters					
Landscaping					

215

SAMPLE SCHEDULE

ITEM	WEEK 5				
	MONDAY	TUESDAY	WEDNESDAY	THURSDAY	FRIDAY
Preconstruction					
Road/Driveway					
Well					
Septic System					
Site Preparation					
Excavation					
Foundation					
Termite Treatment					
Slab					
Framing	Set Logs	Set Logs	Set Logs	Set Logs	Set Logs
Roofing					
HVAC					
Plumbing					
Electric					
Insulation					
Drywall					
Painting					
Trim					
Appliances					
Cabinets					
Floor Coverings					
Gutters					
Landscaping					

Sample Schedule

WEEK 6

ITEM	MONDAY	TUESDAY	WEDNESDAY	THURSDAY	FRIDAY
Preconstruction					
Road/Driveway					
Well					
Septic System					
Site Preparation					
Excavation					
Foundation					
Termite Treatment					
Slab					
Framing	Set 2nd Floor System	Set 2nd Floor System	Set Gable Logs	Set Gable Logs	Frame Dormer
Roofing					
HVAC					
Plumbing	Select Fixtures				
Electric					
Insulation					
Drywall					
Painting					
Trim					
Appliances					
Cabinets					
Floor Coverings					
Gutters					
Landscaping					

SAMPLE SCHEDULE

ITEM	MONDAY	TUESDAY	WEDNESDAY	THURSDAY	FRIDAY
WEEK 7					
Preconstruction					
Road/Driveway					
Well					
Septic System					
Site Preparation					
Excavation					
Foundation					
Termite Treatment					
Slab					
Framing	Frame Roof	Frame Roof	Install Soffit Shingle	Install Soffit Shingle	Install Exterior Trim
Roofing					
HVAC					
Plumbing					
Electric					
Insulation					
Drywall					
Painting			Sand Logs	Finish Ext. Trim	Seal Int. Logs
Trim					
Appliances					
Cabinets					
Floor Coverings					
Gutters					
Landscaping					

SAMPLE SCHEDULE

WEEK 8

ITEM	MONDAY	TUESDAY	WEDNESDAY	THURSDAY	FRIDAY
Preconstruction					
Road/Driveway					
Well					Hook-up Well
Septic System		Install Septic	Install Septic	Install Septic	Hook-up Septic
Site Preparation					
Excavation					
Foundation					
Termite Treatment					
Slab					
Framing	Frame Int. Partitions		Set Windows	Set Exterior Doors	
Roofing					
HVAC		HVAC Rough-in	HVAC Rough-in		
Plumbing		Plumbing Rough-in	Rough-in, Set Tub		
Electric				Inspect Rough-in Electric Rough-in	Electric Rough-in
Insulation					
Drywall					
Painting	Seal Exterior Logs				
Trim					
Appliances					
Cabinets					
Floor Coverings					
Gutters			Install Gutters		
Landscaping					

SAMPLE SCHEDULE

ITEM	WEEK 9				
	MONDAY	TUESDAY	WEDNESDAY	THURSDAY	FRIDAY
Preconstruction					
Road/Driveway					
Well					
Septic System					
Site Preparation					
Excavation					
Foundation					
Termite Treatment					
Slab					
Framing		Framing Inspection			
Roofing					
HVAC					
Plumbing					
Electric	Rough-in, Inspection				
Insulation			Insulate		
Drywall				Deliver Drywall	Hang Drywall
Painting					
Trim					
Appliances					
Cabinets					
Floor Coverings					
Gutters					
Landscaping					

SAMPLE SCHEDULE

ITEM	WEEK 10				
	MONDAY	TUESDAY	WEDNESDAY	THURSDAY	FRIDAY
Preconstruction					
Road/Driveway					
Well					
Septic System					
Site Preparation					
Excavation					
Foundation					
Termite Treatment					
Slab					
Framing					
Roofing					
HVAC					
Plumbing					
Electric					
Insulation					
Drywall	Hang Drywall	Finish Drywall	Finish Drywall	Finish Drywall	Clean-up
Painting					
Trim					
Appliances					
Cabinets					
Floor Coverings					
Gutters					
Landscaping					

SAMPLE SCHEDULE

ITEM	WEEK 11 MONDAY	TUESDAY	WEDNESDAY	THURSDAY	FRIDAY
Preconstruction					
Road/Driveway					
Well					
Septic System					
Site Preparation					
Excavation					
Foundation					Final Grade
Termite Treatment					
Slab					
Framing					
Roofing					
HVAC					
Plumbing					
Electric					
Insulation					
Drywall					
Painting	Paint	Paint	Paint		
Trim				Stain, Finish Trim Install T&G Walls	Stain, Finish Trim Install T&G Ceiling
Appliances					
Cabinets					
Floor Coverings					
Gutters					
Landscaping					

Sample Schedule

ITEM	MONDAY	TUESDAY	WEDNESDAY	THURSDAY	FRIDAY
WEEK 12					
Preconstruction					
Road/Driveway					
Well					
Septic System					
Site Preparation					
Excavation					
Foundation					
Termite Treatment					
Slab					
Framing					
Roofing					
HVAC					
Plumbing					
Electric					
Insulation					
Drywall					
Painting	Stain, Finish Trim	Seal, Finish T&G	Finish T&G		
Trim	Install T&G Walls	Install T&G Ceiling	Hang Doors, Trim	Install Window Trim	Install Closet Trim
Appliances	Deliver Appliances				
Cabinets		Install Cabinets	Install Vanities		
Floor Coverings			Install Vinyl, Ceramic	Install Vinyl, Ceramic	Install Hardwood
Gutters					
Landscaping	Rake, Seed, Straw	Rake, Seed, Straw			

223

SAMPLE SCHEDULE

ITEM	WEEK 13				
	MONDAY	TUESDAY	WEDNESDAY	THURSDAY	FRIDAY
Preconstruction					
Road/Driveway					
Well					
Septic System					
Site Preparation					
Excavation					
Foundation					
Termite Treatment					
Slab					
Framing					
Roofing					
HVAC	Finish HVAC	Finish HVAC			
Plumbing	Install Fixtures	Install Fixtures	Final Inspection		
Electric	Install Switches	Install Plugs	Install Fixtures	Final Inspection	
Insulation					
Drywall					
Painting					
Trim	Install Stairs	Install Stairs			
Appliances					
Cabinets					
Floor Coverings	Install Hardwood			Sand Hardwood	Seal Hardwood
Gutters					
Landscaping					

SAMPLE SCHEDULE

ITEM	WEEK 14				
	MONDAY	TUESDAY	WEDNESDAY	THURSDAY	FRIDAY
Preconstruction					
Road/Driveway					
Well					
Septic System					
Site Preparation					
Excavation					
Foundation					
Termite Treatment					
Slab					
Framing					
Roofing					
HVAC					
Plumbing					
Electric					
Insulation					
Drywall					
Painting			Touch-up	Touch-up	
Trim					
Appliances					
Cabinets					
Floor Coverings	Finish Hardwood	Finish Hardwood	Install Carpet		
Gutters					
Landscaping					FINAL INSPECTION

APPENDIX XII. COST SUMMARY SHEET

BID/ ESTIMATE	DESCRIPTION	COMMENTS
_____	Lot Cost	land, closing costs, taxes, easements
_____ _____	Survey Plans	all plans, architectural fees, engineering fees not included with kit
_____	Permits/Fees	building permit, driveway entrance fee, impact fees, etc. Contact local building authority for complete listing
_____	Sewer/Septic	perk test, septic system installation, install line & hook-up to house
_____	Water Connection/Well	drill well, install pump & tank, hook up to house, water test if required

_____	Electric Service	install power line, transformer, set meter & connect service
		    to house electric panel; contact local power company
		    for estimate

_____ _____	Portable Toilet Storage Van	
_____	Lot Layout	surveyor or contractor to mark house corners & set offset stakes for digging foundation
_____	Clearing/Site Preparation	clear trees & underbrush, cut wood into firewood lengths, stack, remove or bury stumps

Cost Summary Sheet (continued)

BID/ESTIMATE	DESCRIPTION	COMMENTS
___	Excavating	dig foundation, separate topsoil
___	Footings	dig footings, install steel, pour concrete
___	Foundation	erect foundation walls, install drain tile around foundation, apply waterproofing
___	Steel	furnish & install any special steel reinforcement or beams required
___	Termite Treatment	treat foundation before backfilling, furnish certificate & warranty
___	Backfill/Rough Grading	Backfill around foundation & smooth to within six inches of final grade
___	Gravel Fill	furnish & install, if required
___	Log Package, including delivery	Complete delivered kit cost
___	Unloading Logs inc. forklift, labor	forklift rental, labor to unload & stack kit (may be included with construction contract)
___	Windows/Doors not in pkg	delivered price
___	Framing Materials not in pkg	delivered price for framing not in kit
___	Framing Labor	cost to erect all or part of kit
___	Masonry/Stone	material & labor for stone veneer on foundation, retaining walls, fireplace chimneys
___	Roofing Material not in pkg	shingles, rafters, beams not in kit

Cost Summary Sheet (continued)

BID/ ESTIMATE	DESCRIPTION	COMMENTS
_____	Roofing Labor	shingling including drip edge & ridge vent
_____	Concrete Flatwork	basement, garage, porch floor slabs
_____	Exterior Paint/Log Sealant	materials & labor
_____	Garage Doors/Openers	materials & installation labor
_____	Rain Gutters	materials & labor, include splashblocks for downspouts & outside water spigots
_____	Decks not in pkg	materials & labor
_____	Insulation	materials & labor for ceiling, roof, subfloor or foundation, framed wall insulation, check local building code requirements.
_____	Drywall	furnish, install, finish joints
_____	Interior Finish Material	baseboards, closet shelves, rods, crown & specialty mouldings, interior doors, jambs, trim, interior window trim, stairs
_____	Interior Finish Labor	materials & labor, including sanding, filling nail holes
_____	Interior Painting/Log Finish	materials & labor, include underlayment, adhesives, grout
_____	Ceramic Tile	materials & labor, stair & loft rails
_____	Wrought Iron/Wood Railing	kitchen, bath, laundry/utility room cabinets, vanities, medicine cabinets
_____	Cabinets/Vanities/Countertops	materials & installation, include underlayment, adhesives
_____	Vinyl Flooring	

Cost Summary Sheet (continued)

BID/ ESTIMATE	DESCRIPTION	COMMENTS
_____	Hardwood Flooring/inc. floor finish	materials & labor, flooring, underlayment, sand, seal, finish
_____	Carpet	materials & labor
_____	Caulking/Dampproofing	materials & labor
_____	Appliances	refrigerator, range, microwave, cook-top, dishwasher, disposal, washer, dryer
_____	Finish Hardware	door locksets, hinges, door stops, towel bars, toilet paper holders
_____	Fireplaces/Stoves	material & installation, fireplace, chimney, hearth, mantel, cap
_____	Electrical	install service panel, rough wiring, switches, plugs, covers, hook up appliances, HVAC, hot water heater, smoke detectors, cable TV, telephone
_____	Electrical Fixtures	material & labor for lighting fixtures, flood lights, motion sensors
_____	HVAC	materials & labor for heating, ventilation & air conditioning, include bathroom, dryer vents, humidifiers, dehumidifiers, air cleaners
_____	Plumbing	materials & labor for rough plumbing, hot water heaters, tubs, showers, toilets, bidets, saunas, steam rooms, hook up ice-maker, disposal, sump pump

229

Cost Summary Sheet (continued)

BID/ ESTIMATE	DESCRIPTION	COMMENTS
_____	Telephone Wiring	if separate from electric
_____	TV Prewire	if separate from electric
_____	Final Grade	replace topsoil
_____	Rake, Straw, Seed	materials & labor
_____	Landscape	materials & labor as per landscape plan
_____	Drives, Walks, Patios	materials & labor to clear, grade, gravel & finish entrance road or drive, including, install drainage pipes, railings, asphalt or concrete if required
_____	Retaining Walls	materials & labor
_____	Trash Removal, Clean-up	removal of construction debris, include hauling & dumping fees
_____	**Subtotal**	
_____	Financial Costs	obtain estimate from lender
_____	**Total**	add minimum of 5% contingency for unexpected expenses

Appendix XIII. Kit Comparison Sheet

MATERIALS INCLUDED IN KIT:	MANUFACTURER				
	Log Homes	Log Homes	Log Homes	Log Homes	Log Homes
Sill Sealer Insulation					
CCA Treated Sill Plates					
Subfloor Adhesive					
Metal Floor Joist Bridging					
Basement Support Posts					
First Floor Girder Beam					
First Floor Floor Joists — 16″ o.c.					
First Floor ¾″ T & G Plywood					
Second Floor 6 x 8 Oak Beams					
Second Floor Oak Girder Beams					
Laminated Girder Beams if req'd					
Beam Hangers					

Kit Comparison Sheet (continued)

MATERIALS INCLUDED IN KIT:	MANUFACTURER					
	Log Homes	Log Homes	Log Homes	Log Homes	Log Homes	Log Homes
Support Posts for Second Story						
Second Story Flooring						
Partition Wall Framing						
Gable Framing, or Log Gables						
Pre-fab Roof Trusses if req'd						
Roof Rafters & Ridge Poles if req'd						
Collar Beams or Ceiling Joists if req'd						
Laminated Roof Beams if req'd						
Exposed Beam Trusses if req'd						
Struts and/or Jack Rafters if req'd						
½" Plywood Roof Sheathing						
Plywood Clips for Truss Roof						
Shingles						
Starter Shingles & Ridge Caps						

Kit Comparison Sheet (continued)

MATERIALS INCLUDED IN KIT:	MANUFACTURER				
	Log Homes	Log Homes	Log Homes	Log Homes	Log Homes
Roofing Felt					
Roofing Nails					
Truss/Snow Blocking as req'd					
Fly Rafters & Lookout Framing					
Wall Plates & Overhang Framing					
Truss/Rafter Bands					
Cedar Fascia Boards					
Metal Roof Edge					
Soffit Enclosure					
Ridge or Gable Vents					
Soffit Vents					
Sheathing for Gables (if not log)					
Gable End Siding (if not log)					
Nails					

Kit Comparison Sheet (continued)

MATERIALS INCLUDED IN KIT:	MANUFACTURER				
	Log Homes	Log Homes	Log Homes	Log Homes	Log Homes
Basement Steps					
Second Story Steps					
Spiral Stairs, if shown					
Exterior Doors					
Sliding, French, or Swinging Exterior Doors					
Interior Doors					
Closet Bi-Fold Doors					
Garage Doors if req'd including installation					
Door Jambs					
Interior & Exterior Door Trim (cedar)					
Windows Screens & Grilles					
Skylights if shown					
Circle-top Windows if shown					
Trapezoid or other custom shape if shown					

Kit Comparison Sheet (continued)

MATERIALS INCLUDED IN KIT:	MANUFACTURER				
	Log Homes	Log Homes	Log Homes	Log Homes	Log Homes
Window Jambs					
Interior & Exterior Window Trim					
Dormer Framing if req'd					
Dormer Sheathing					
Dormer Siding					
Dormer Roof, Soffits, etc. as in Roof					
Treated Porch Girders & Floor Joists					
Porch Posts					
Porch Flooring					
Porch Rafters					
Porch Ceiling Joists					
Support Beams for Porch					
Porch Ceiling					
Garage Carport Roof System if req'd (trusses, plywood, shingles, etc.)					

Kit Comparison Sheet (continued)

MATERIALS INCLUDED IN KIT:	MANUFACTURER				
	Log Homes	Log Homes	Log Homes	Log Homes	Log Homes
Garage Carport Ceiling					
Garage Carport Posts & Support Beams					
Log Walls					
Logs or Siding to encircle 1st story subfloor					
Logs or Siding to encircle 2nd story subfloor					
Full Log Gable Ends for two story & loft models					
Full Log Gambrel Ends where req'd					
Delivery, specify					
Spikes					
Log Sealant					
Caulk					
Blueprints					
Construction Manual					
Ceiling/Wall Decking					

Kit Comparison Sheet (continued)

MATERIALS INCLUDED IN KIT:	MANUFACTURER				
	Log Homes	Log Homes	Log Homes	Log Homes	Log Homes
Stair Railings					
Treated or Redwood Deck					
Porch Railings					
Exterior Wood Sealant					
Hardwood Flooring					
Storm Doors					

GLOSSARY

Allowance. A specified dollar amount included in the contract for items that are highly variable in cost, such as cabinets, carpet, ceramic tile, lighting fixtures, etc. The contractor bases his contract price on a certain amount specified in the contract as an allowance. If the cost of the item selected by the home owner exceeds the allowed amount, the homeowner must pay the difference. If the item costs less than the amount stated in the contract, the homeowner receives a credit.

Backfill. Replacement of earth against the foundation in the trench left by excavating the foundation.

Bearing points or walls. The points such as posts or walls that receive and carry the weight of the house and roof structure.

Borate. A type of chemical insect treatment based on borax or sodium borate. Currently popular for its effectiveness while showing very low toxicity to pets and people.

Builder. An individual or company that employs construction workers as opposed to using subcontractors. Most builders still must subcontract certain portions of a complete log home.

Builder's level. An instrument consisting of a low power telescope with cross hairs and bubble levels built into its base. Used to measure elevation and check levelness of foundation walls, footings, and excavations.

Builder's risk insurance. A special policy that insures material and labor investments during the construction process.

Calcium. Used in the construction industry as an additive to concrete to aid drying in cold weather.

Checking. Separation of wood fibers in a log as it dries that appears as a crack in the log. Also called season-checking because it occurs as the wood seasons. Logs and large timbers of virtually all woods can check. The amount, size, and depth of checking varies with the wood species. Checks are a natural part of the seasoning process of the log and generally do not affect its structural integrity or weathertightness.

Chinking. Originally a variety of materials such as wood, mud, mortar, or cement used to seal the joint between two logs. Characteristic of log styles in which logs do not rest directly on each other. Modern chinking is usually a mortar or caulk that is flexible enough to permit some movement of logs.

Clear title. A real estate term indicating that ownership of real property has been clearly established and all costs related to ownership have been paid.

Cold joint. An irregular seam in a concrete floor or wall that results when poured concrete dries to the extent that additional fresh concrete will not mix with it. Usually occurs when too much time lapses between truckloads of concrete.

Come-along. A device consisting of one or two cables and a pulley and gear system. Used to pull large objects together, such as in straightening log walls and tightening beams.

Compression pack. A spring that is placed under the head of a spike before it is driven into a log. Acts to exert a steady pressure on the log as it settles.

Construction calculator. A handy device similar to a regular pocket calculator except that it functions in units of feet and inches or meters. Also contains internal functions for calculating rafter lengths, stair dimensions, roof pitches, etc.

Contingency. A clause in a contract that modifies the contract in specific situations. An example is a "well" contingency in a land contract, which says that the purchaser has the right to withdraw from the contract with no penalty if the property fails to support a well that produces sufficient water and meets health requirements.

Crawlspace. A shallow space beneath a house enclosed by a foundation wall, but with no floor.

Critical path. A method of scheduling built around a sequence of construction stages that must be completed before the next stage can be undertaken.

Crown. A natural bow or arch that often occurs in dimensional lumber, beams, or even logs. The face or edge of the lumber that arches upward is said to be "crowned." When setting subfloors, rafters, and beams, crowns should be upward to counteract downward forces that might cause sagging.

Crown Mold. A decorative molding placed at the joint between wall and ceiling or to conceal a settling space located near the ceiling.

Deck. An exposed floor without a roof attached to the house. Also a term for first-story subfloor.

Diagonal. A measurement taken between opposite alternate corners of a rectangle or square. Diagonals break a rectangle into two triangles. Used to check squareness. If both diagonals of a rectangle do not have the same measurement, it is "out of square."

Dimensional lumber. Kiln-dried smooth-surfaced lumber available in 2-inch width increments and 1-inch thickness increments up to 4 inches thick. Used for joists, studs, rafters, and other framing.

DIY. An abbreviation for "do-it-yourself."

Dovetail. A wedge-shaped joint in logs or cabinetry. Its shape forces the parts of the joint together, adding strength.

Drain tile. Perforated plastic pipe that surrounds the footings of a foundation, used to carry water away from the foundation.

Dried-in shell. The outer shell of the house completed to a stage that prevents entry of rain or wind. Definitions vary; I use it to mean completion of walls and roof, shingles installed, exterior doors and windows installed and trimmed on the outside, dormers sheathed, and soffit and fascia installed.

Drywall. Also called gypsum board. A sheet product made by enclosing gypsum in a paper covering.

Easements. Legal agreements affecting land contracts that specify certain exceptions to ownership rights. For example, most property sold today contains utility easements that allow utility companies access to a narrow strip along the property edge to install and maintain utility lines. In cases where property does not have direct access to public roads, an entrance easement from a neighbor may be necessary.

Fascia. A board that surrounds the end of the roof, concealing framing.

Finish schedule. Part of a building contract, specification sheet, or blueprint set that states how specific areas of the house will be painted, stained, sealed, etc.

Firm bid. A guaranteed price submitted by a subcontractor or supplier to provide certain materials or to perform specific work. The exact nature of the materials or work is specified in the bid. Firm bids generally are submitted with specific limits (usually 30–90 days), at which time the guarantee expires.

Flashing. Metal or vinyl used to keep water from entering joints between construction components. For example, flashing around a chimney prevents water from running down the chimney into the roof.

Fly rafter. The end rafter on a roof overhang that is supported by the ridge board and lookouts, short pieces that extend out from the wall of the house.

Footer or Footing. A concrete or gravel base upon which the foundation walls of a house rest.

Forklift. A machine with extensions that elevate on its front, capable of lifting and transporting heavy loads. Rough-terrain forklifts have heavy tires and are used on jobsites.

Frost footer. A footer near the surface of the ground that extends down below the penetration of frost. Depth varies in different parts of the country but frost footers are necessary where the base of a foundation is exposed, as in a walk-out basement.

Gable. The vertical triangular portion of an end wall of a house beneath the roof.

General contractor. A construction professional who relies on subcontractors rather than using paid employees to construct a house.

Grade beam. A concrete beam formed beneath a concrete floor by digging a trench and sometimes adding lengths of steel. Used beneath garage slabs or basement floor slabs that rest on fill dirt.

Handcrafter. A log home builder who constructs a house using raw logs and lumber, cutting and fitting individual pieces.

Hip. An external angle formed where two sloping sides of a roof intersect at an angle to the direction of the walls.

Kiln-drying. A process for removing moisture from lumber under controlled temperature and humidity conditions. Done to reduce shrinkage, twisting, and checking. Dimensional lumber used in home construction is kiln-dried. Large timbers cannot be kiln-dried to the same moisture levels as dimensional lumber.

Lag screw. A large diameter screw with a square or hexagonal head to allow tightening with a wrench or ratchet.

Lead time. The amount of advance notice that suppliers and subcontractors need to arrive at your job when their services will be required.

Mechanical system or "mechanicals." Includes plumbing, electrical wiring, and HVAC (heating, ventilation, and cooling).

Nailbase. A plywood or composition base into which shingles can be nailed when using solid foam roof insulation.

Offsets. Stakes placed by a surveyor and builder locating the corners of a building. The stakes are set back from the actual corners to allow the excavator to work and to allow location of corners when excavation is complete.

Parging. A coating of cement applied to the surface of masonry block foundations. Not the same as waterproofing.

Partition. A wall that subdivides any portion of a house.

Perk test. An official test conducted by health officials to determine the suitability of soil for a septic system. Conducted by digging a hole in an area designated for a septic field, filling it with water to a certain depth, and then measuring the time required for the hole to drain. A requirement in many areas for building a new home where a septic system must be used.

Plat. A map prepared by a surveyor, showing property boundaries, easements, and legal features.

Plumb. Exactly vertical or at right angles to level.

Plumbing wall. A wall containing water, waste, and vent pipes. Usually framed of 2″ x 6″ dimensional lumber.

Portland cement. A component of concrete. Used alone to form a hard surface, as in parging for masonry block walls.

Precut log kit. A log kit in which the manufacturer cuts logs to exact lengths, assembles, numbers, and disassembles them before shipping them to the site where they will be erected.

Preperk test. Identical to a perk test except done unofficially, usually for a prospective property owner before finalizing the purchase of land. Preperk tests can be done by any septic system installer.

Prequalify. An evaluation of a prospective home buyer's financial condition to determine the amount of permanent financing for which he or she is eligible.

Purlin. Logs or heavy timbers running horizontally and supporting rafters.

Radon. A naturally occurring gas, found in certain geological conditions, that can accumulate in basements. It has been implicated in causing lung cancer.

Radon loop. A coil of plastic pipe similar to drain tile, lying beneath a basement floor slab and terminating outside the house or in a specially covered sump crock in the basement. Air can be circulated beneath the slab and exhausted to the outside of the house if high levels of radon are detected in the basement.

Ridge vent. A metal cap or strip of mesh concealed beneath shingles that allows hot air to escape from the roof.

Rock clause. A specific clause in most builders' or excavators' contracts specifying that their price does not include nonstandard excavation techniques such as blasting, should heavy rock be encountered. Even when not written, in many areas it is accepted as implied.

Rough-in. Basic wiring, plumbing, or ductwork to which fixtures will be attached to complete mechanical systems.

Sand mound septic system. A special septic system used in some areas where soil conditions will not support a conventional septic system. The septic field is created by importing special sand through which the drain field will run.

Septic system. An alternative system for removal of waste in areas where sewers are not available. Consists of a septic tank that receives waste and a drainage field containing trenches or a pit.

Setback. A legally specified distance that a structure (e.g., road or house) must maintain from property boundaries, easements, utilities, or other features.

Settlement space. A space between logs or within framing, designed to allow for the tendency for logs to settle. Usually found above windows and exterior doors. Also at the top of interior partition framing.

Sewage injector pump. A mechanical device that pumps sewage from a storage area beneath a basement floor into the waste system of the house. Used where gravity flow is not possible.

Shim. A small piece of metal, plastic, or wood used to separate, support, or align components.

Sill plate. A piece of dimensional lumber, treated for moisture protection that sits on top of the foundation wall and acts as the base for subfloor and house walls.

Sill sealer insulation. A flat strip of foam that lies underneath the sill plate.

Silt fence. A device consisting of a heavy fabric attached to wooden stakes. Used to prevent soil from disturbed areas such as excavations from getting into waterways, roads, and other properties.

Site plan. A drawing prepared by the homeowner, architect, landscape architect, or surveyor that shows the position of the house, well, septic system, and road on the property.

Soffit. The covering for the underside of roof overhangs.

Soffit vent. Perforated metal, mesh, or screening installed in the soffit to allow circulation of air beneath the roof sheathing.

Splash blocks. Masonry, stone or fiberglas blocks that receive water exiting the downspout and divert it away from the house.

Stress skin panels. Structural panels consisting of a core of insulation with a nailbase such as plywood, chipboard, or finished surfaces on outside and inside.

Strongback. A vertical timber or framing installed to add support to long runs of unbroken log wall.

Subfloor. The floor joists and plywood sheathing that support finished flooring and framing.

Sump crock. A container set beneath the basement floor that collects water when drain tile cannot provide gravity flow to the ground surface away from the house. A sump pump installed in the crock automatically starts when water reaches a certain level, pumping water to the ground surface outside the house.

Sweat equity. The contribution of labor by a home owner toward completion of a log home.

Take-off. A detailed analysis of blueprints used to produce a list of all materials required to complete a house.

Tap fee. A fee paid to hook into a public water or sewer system.

Termite shield. A metal strip that lies atop the foundation and extends outward from the wall to discourage termites.

Through-bolt. A log-wall fastening system consisting of threaded rods extending through the entire log wall from top to bottom and bolted at both ends. Periodic tightening of the bolt located beneath the subfloor takes up any settlement.

Title insurance. Insurance required by lenders in some instances to protect land should problems arise in establishing positive ownership.

Title search. A search of records conducted by a lender to establish ownership of real property.

Torpedo level. A small level, tapered at both ends, useful for checking for level and plumb.

Trades. Refers to the building trades such as masonry, excavation, plumbing, wiring, carpentry.

Transit. Similar to a builder's level but also capable of measuring horizontal angles. See **Builder's level**.

Truss roof. A roof system erected by installing premanufactured triangular framing units called trusses instead of conventional rafters.

Turnkey. Used by builders to denote a complete home. Exact definitions vary. As used here, the house is complete except for furniture, clothing, and groceries.

Underlayment. A layer of material such as plywood or fiberboard placed on top of the subfloor to provide a surface for finished flooring.

Valley. An internal angle formed by the intersection of two sloping sections of a roof.

Walk-out basement. A basement with an entire vertical section of wall above the ground surface. Occurs when houses are built on sloping lots.

Water level. A leveling device consisting of two graduated columns joined by a length of plastic tubing or hose and filled with fluid. Works by setting the base of the columns on points to be checked and comparing the heights of the fluid in the columns. Useful in situations where the points are widely separated and obstacles prevent the use of a builder's level.

Window buck. Dimensional framing that surrounds the log opening of a window into which the actual window unit is set.

Wire mold. Covering that conceals wiring run along the surface of a wall.

Wood foundation. An alternative foundation wall that looks like a conventional framed wall covered with plywood sheathing. All wooden components, however, are heavily pressure-treated with preservatives (much more than that used to treat deck lumber). Wood foundations are usually placed on a gravel base rather than concrete footings.